A-Z OF FISHERIES AND AQUACULTURE TECHNOLOGY

A-Z OF FISHERIES AND AQUACULTURE TECHNOLOGY

K.D. Bhardwaj

CYBER TECH PUBLICATIONS
4264/3, Ansari Road, Daryaganj, New Delhi-110002 (India)
Ph.: 011-23244078, 011-43559448 Fax: 011-23280028
E-Mail: cyberpublicationsdelhi@yahoo.com
Website: www.cybertechpublications.com

A-Z OF FISHERIES AND AQUACULTURE TECHNOLOGY

K.D. Bhardwaj

First Edition 2011

Published by :

G.S. Rawat for **Cyber Tech Publications**
4264/3, Ansari Road, Daryaganj, New Delhi-110002 (India)
Ph.: 011-23244078, 011-43559448 Fax: 011-23280028
E-Mail: cyberpublicationsdelhi@yahoo.com
Website: www.cybertechpublications.com

CONTENTS

Preface

Fisheries is a sunrise sector of our economy. Its role in increasing food supply, generating job opportunities, raising nutritional level and earning foreign exchange has been important. Growing urbanization, globalization, rapidly changing social structures had a major impact on the fisheries structure in the country. Fisheries and aquaculture emerged as the important commercial activity from its traditional role as subsistence supplementary activity.

The publication provides an overview of socioeconomic profile of stakeholders (fishermen/women), fishing and aquaculture technologies, demand and expected supply, policies, institutions and support systems. I hope this volume would be very helpful to policy makers, research managers and others to understand and plan fisheries sector better.

1

Introduction

THE GLOBAL DEMAND FOR FISHERY PRODUCTS

The oceans have historically been seen—and exploited—as an inexhaustible supply of food for human use. This perception, coupled with the concept of open access to that common resource pool, set the stage for ecological disaster. That disaster—in the form of a general collapse of commercial marine fisheries—either looms or has already become reality (see, for example, Pauly et al. 2002, 2003; Myers and Worm 2003; Caddy and Surette 2005; Worm et al. 2006). Persistent fisheries stock declines since the 1980s have been caused by a combination of factors, including increased industrial-scale fishing effort, advances in fishing technology, destructive fishing practices, and fishery management policies dictated by short-term economic interests. The result is that the number of fully exploited or overexploited stocks of marine fish is high and increasing, and the global potential for marine capture fisheries has been reached (FAO 2004c, 2006b).

Marine fisheries catch increased rapidly from 1950, reaching a peak of 80 to 85 million tonnes/year in the late 1980s. Since then, catch has been steady or even declining when corrected for presumed overreporting by China (Watson and Pauly 2001). Inland capture fisheries add about 8 to 9 million tonnes annually, bringing the total world catch from capture fisheries to about 90 million tonnes in 2003. About 30 million tonnes of the 2003 world fisheries catch was destined for nonfood uses—primarily reduction to fishmeal and fish oil for use in animal feeds. Therefore, approximately 60 million tonnes of fishery products destined for human consumption were extracted from world capture fisheries in 2003 (FAO 2004c).

Over the period 1990 to 2003, when foodfish output from capture fisheries was, at best, stagnant, global consumption of edible fishery products tripled to nearly 105 million tonnes annually. The difference between the nonexpanding supply from capture fisheries and the rapidly expanding demand was derived from aquaculture—farming aquatic plants and animals in oceans and inland waters. In 1980, world aquaculture production (excluding plants) was approximately 5 million tonnes, which was approximately 7% of total world foodfish supply. In 2004, nearly 46 million tonnes of fish and shellfish were produced in aquaculture (FAO 2006b), representing almost half of global foodfish production. The annual rate of increase in aquaculture production since 1980 has been approximately 10%, which is faster than that for any other animal food producing sector (FAO 2006b).

Population growth, rising per capita incomes, urbanization, and increased appreciation of the role of seafood in human health will continue to increase the global demand for seafood. Capture fisheries must provide a large part of the world's supply of fish and shellfish, but dramatic changes are needed to assure that marine resources are managed sustainably. Oceans must be protected from environmental degradation caused by pollu-tion and global climate change, and marine fisheries must be managed intelligently to restore and maintain the oceans' biodiversity and ecological integrity. Current "best-case" scenarios for fisheries management indicate, however, that it will not be possible to increase marine fisheries landings past levels obtained in the 1980s (Pauly et al. 2003). Aquaculture must therefore continue to expand to meet any increase in demand for fishery products (FAO 2004a).

The Changing Face of Aquaculture

Aquaculture evolved thousand of years ago as an activity with origins and goals similar to other animal husbandry activities. That is, methods were developed to provide animal protein when local human population growth or overexploitation of accessible wild popu-lations made it difficult to obtain food by hunting—or fishing, in the context of aquacul-ture. Simple but elegant fish culture systems were developed in Asia that in many ways resembled natural aquatic systems in their fundamental ecosystem dynamics. For thou-sands of years aquaculture was practiced as a relatively low-input activity and was seen as a beneficial or, at worst, benign endeavor that provided high-quality animal protein for families or local communities. Much of current aquaculture remains rooted in these ancient practices. Countries in Asia and the western Pacifi c region presently account for approxi-mately 90% of the world aquaculture production, and much of that activity is based on pond aquaculture of grass carp (*Ctenopharyngodon idella*), silver carp (*Hypophthalmichthys molotrix*), bighead carp (*Aristichthys nobilis*), and common carp (*Cyprinus carpio*)—the same species raised in traditional pond aquaculture in China thousands of years ago. In fact, freshwater pond culture of cyprinids for local consumption accounts for almost a third of global aquaculture production, including plants (FAO 2006b).

As aquaculture production rapidly increased in the last half of the 20th century, culture methods and technologies evolved in response to profit incentives and encouragement from aquaculture development agencies. In many instances, the goal of supplying food for local consumption changed to that of producing higher-value products for export. Associated with this trend was the use of culture practices with higher rates of resource use and greater environmental impacts than traditional aquaculture methods. The rapid expansion of aquaculture development occurred during a time of heightened environmen-tal awareness and advocacy, and well-publicized problems in certain sectors led to closer scrutiny of aquaculture in general.

Based on real and perceived environmental impacts, some critics believe that aquaculture cannot—and should not—meet the increasing global demand for seafood. In one respect this argument is moot because capture fisheries clearly cannot grow to meet the demand. More importantly, critics of aquaculture often fail to see that aquaculture is a diverse endeavor and that a monolithic "aquaculture industry" does not exist. Culture practices and, correspondingly, environmental impacts vary widely. Further, although aquaculture is an ancient practice, little thought has been given—until recently—to the environmental impacts of aquaculture and to ways in which those impacts can be reduced. In that regard, commercial aquaculturists and aquaculture scientists are indebted to envi-ronmental groups and forward-looking members of the aquaculture industry for pointing out problems associated with certain aquaculture practices. The hope and expectation that aquaculture will meet the long-term shortfall in seafood production will be fulfilled by acknowledging and addressing the need for improved environmental performance.

Aquaculture production can, and must, increase over the foreseeable future and this expansion must be conducted in a responsible manner. Development must balance a number of difficult and, at times, conflicting issues and considerations, including consumer food preferences; short- and long-term economic benefits of aquaculture development; the role of aquaculture in rural development, trade balance, and food-supply biosecurity; and the capacity of the local, regional, and global ecosystems to support aquaculture produc-tion. Some issues will be difficult to address. For example, a common criticism of aqua-culture involves the farming of high-value, carnivorous species that require considerable ecosystem support to produce. Addressing this concern will require nothing less than changing the social values of consumers in developed countries who are willing to pay relatively high prices for those products. On the other hand, many of the environmental problems ascribed to aquaculture can be addressed by changing or improving culture methods. The range of environmental impacts must be identifi ed, and the aquaculture community must address problems by changing practices or developing new technologies to mitigate negative impacts. Significant improvements have been made in resource use efficiency and environmental performance of aquaculture, and further improvement is possible.

AQUACULTURE IN THE UNITED STATES

About 45.5 million tonnes of fish and shellfish were produced in aquaculture in 2004. Of that, 31 billion tonnes—nearly 70%—were produced in China. According to FAO (2006c), approximately 600,000 tonnes of fish and shellfish were produced in United States aqua-culture in 2004. This is approximately 1.3% of global production, ranking the United States tenth in global fish and shellfish aquaculture. When China is excluded, United States aquaculture accounted for 4% of world production in 2004.

Despite the relatively small contribution to global production, aquaculture plays a significant role in United States trade and agriculture, and there is considerable incentive for further development. Per capita seafood consumption in the United States was 21 kg live weight equivalent in 2004 (NMFS 2005), which was slightly greater than the global average of about 16 kg live weight equivalent. This, coupled with a large population, makes the United States the third largest consumer of edible fisheries products in the world. Although the United States is one of the largest exporters of seafood products, it is second only to Japan as the largest importer of fishery commodities (FAO 2004b), resulting in a significant international trade defi cit. In 2004, the United States imported $11.2 billion dollars of fish and shellfish products while exporting $3.8 billion, for a trade defi cit of approximately $7.4 billion, which increased to more than $8.8 billion in 2006. In 2006 the trade defi cit in fish and shellfish products was the 24th largest trade defi cit item for all commodity groups and was the largest defi cit item for any agricultural commodity. As with the global situation, United States seafood demand will continue to increase as a result of population growth and increased emphasis on eating seafood as part of a healthy diet. Assuming that desirable local, regional, and national economic and food security benefits accrue from producing fishery products rather than importing them, domestic aquaculture production should grow to meet the increasing demand for seafood by consumers.

Production

The United States has a varied geography and climate, providing opportunities for a diverse aquaculture industry. Physical resources in the United States make it possible to raise fresh-, brackish-, or saltwater organisms from tropical, temperate, and arctic climates. At least 50 aquatic animal species are grown in commercial scale (USDA-NASS 2006), and probably at least that number again are raised for local use, recreation, developmental aquaculture, or research. Reported sales of aquaculture products in the United States exceeded $1 billion in 2005 (USDA-NASS 2006). Estimates of aquaculture production by weight vary among the various data-collecting organizations and range from about 450,000 to more than 600,000 tonnes for the 2003–2005 period (NMFS 2005; FAO 2006a, b, c; USDA-NASS 2006). Variation in reported aquaculture production depends on the basis for reporting shellfish harvest weights (live weight or meat weight only) and the extent to which species produced in "miscellaneous" aquaculture are included in the database.

Although many aquatic species are grown in the United States, production is skewed toward freshwater and a relatively small number of species. Based on USDA-NASS (2006) figures, freshwater aquaculture contributes about 70% of total United States aquaculture production by weight and about 60% by value. One fish—channel catfish (*Ictalurus punctatus*)—accounts for almost 60% of total production by weight and nearly 45% by value.

Aquaculture production was reported from every state in 2005 (USDA-NASS 2006). Production is, however, unevenly distributed. The four deep-south states of Alabama, Arkansas, Louisiana, and Mississippi account for almost 60% of United States aquaculture production value. Channel catfish is the primary species raised in Alabama and Mississippi, and catfish and baitfish are raised in Arkansas. Louisiana has a diverse aquaculture industry with significant production of crawfish, oysters, alligators, catfish, pet turtles, and baitfish. Idaho, Washington, California, and North Carolina lead the country in trout production. Florida has a diverse aquaculture industry and is unique because the most important crop is freshwater ornamental fish, with sales exceeding $40 million in 2005.

Washington state is the fifth largest aquaculture-producing state, with well-established molluscan shellfish and Atlantic salmon farming. Other states with significant marine aquaculture include Maine, which leads the country in Atlantic salmon production, and California, Connecticut, Florida, Louisiana, Massachusetts, Oregon, and Virginia—all of which have significant shellfish aquaculture industries.

Additional information on the major species grown in United States aquaculture is found in Chapters 6 through 12. Detailed production data and summaries of United States aqua-culture are available in various FAO publications (FAO 2004b, c; 2006a, b, c), the United States Department of Agriculture's Census of Aquaculture (USDA-NASS 2006), the National Marine Fisheries Service's report on Fisheries of the United States (NMFS 2005), and the report on marine aquaculture prepared by the Marine Aquaculture Task Force (2007).

Facilities

The goal of this book is to provide technical guidance to improve environmental perfor-mance of aquaculture in the United States. This presentation could be arranged by species or culture system because specific impacts vary with both factors. On balance, however, many environmental issues, such as pollution, water use, predator control, and escapes are more strongly related to culture system than to species, and that is the basis for the organization of this book.

There were 4,309 aquaculture farms operating in the United States in 2005 (USDA-NASS 2006). Most farms were located in the southeastern United States, led by Louisiana with 873 farms, Mississippi with 403, Florida with 359, Alabama with 215, and Arkansas with 211. The two states outside the southeast with the highest number of aquaculture farms were Washington (194) and California (118). Although a wide variety of production facilities is used in United States aquaculture, ponds are by far the most commonly used culture system.

Ponds

Aquaculture ponds are confined bodies of standing water managed to produce a crop of finfish or shellfish. Ponds may be filled with fresh-, brackish-, or saltwater, and are usually constructed of soil, although some may be lined with plastic or other materials to reduce seepage. An important part of the definition of ponds is the long hydraulic residence time, which allows much of the waste produced during culture to be removed before water is discharged. In that regard, ponds are similar to water recirculating aquaculture systems, although water is treated in distinct unit processes in recirculating systems whereas waste treatment and aeration are inherent in the pond ecosystem.

Approximately 27 million tonnes of fish and crustaceans were produced in ponds throughout the world in 2003, representing about 70% of global aquaculture production, excluding plants (FAO 2006c). In the same year, more than 310,000 metric tonnes of fish, crawfish, and marine shrimp were produced in 140,000 ha of ponds in the United States, which was about 65% of United States aquaculture production (USDA-NASS 2006). Almost 90% of the finfish produced in the United States was grown in ponds and more than 99% of the ponds are inland, freshwater ponds used primarily to grow channel catfish, crawfish, bait and ornamental fish, and hybrid striped bass. A few hundred hectares of ponds located on the Texas coast are used to produce marine shrimp.

Net pens

Net pens and cages (collectively referred to here as net pens) are submerged, suspended, or floating confinement systems in which aquatic animals are grown. Net pens may be located along a shore or pier or anchored and floating offshore, either in freshwater or saltwater. Net pens rely on tides, currents, and other natural water movements to supply oxygenated, high-quality water to farmed animals. Farms are often located in public waters leased from the government.

The use of net pens for aquaculture has grown rapidly since 1990, primarily as a result of the widespread interest in growing salmonids and other marine fish. Nevertheless, only about 15% of global aquaculture production by weight was derived from net pens and cages in 2003, although this represented almost 30% of production by value because net pens are primarily used to grow high-value species (FAO 2006b). In the United States, net pens are used almost exclusively to grow Atlantic salmon (*Salmo salar*), which in 2003 accounted for about 2% of domestic production by weight and 4% by value.

Flow-through systems

Flow-through systems, as the name implies, are aquaculture systems with continuous water inflow and outflow. Suitable water quality for aquatic animal culture is maintained by the steady supply of oxygen in the incoming water and removal of waste products in the outflow. Culture units in flow-through systems include earthen or concrete raceways, or tanks constructed from a variety of materials. Culture units can be arranged so that water passes through each unit once without reuse or in a series where water flows by gravity from one culture unit to another. When operated in a series, water is reaerated and, pos-sibly, treated to remove solids or other waste products before reuse in the next unit down-stream. The most common water sources for commercial flow-through systems are artesian springs or water diverted from streams or rivers. Rarely, water can be pumped from a water body, allowed to flow through the raceways or tanks, and then returned to the original body of water.

Flow-through systems are most commonly used to produce trout and other salmonids. Some warmwater species are also produced in flow-through systems, but to a much smaller extent than salmonids. Approximately 28,000 tonnes of fish, representing about 6% of total United States aquaculture production, were produced in flow-through systems in 2003.

Recirculating aquaculture systems

Recirculating aquaculture systems consist of a culture unit connected to a set of water-treatment units that allows some of the water leaving the culture unit to be reconditioned and reused in the same culture unit. Recirculating aquaculture systems minimally require water treatment processes to remove solids, remove or transform nitrogenous wastes, and add oxygen to the water. Other processes—such as temperature control, pH adjustment, gas removal, and disinfection—may also be required. Recirculating aquaculture systems minimize water use and effl uent volume, and generally allow for greater control of the culture environment than is possible in other systems. However, the costs associated with construction and operation can increase the cost of producing fish. Commercial recirculat-ing aquaculture systems are therefore used to produce relatively higher-value fish or in public facilities to produce fish for recreational stock enhancement or restoration of threatened and endangered species.

More than 75% of the annual United States tilapia production and about a third of the food-sized hybrid striped bass are grown in recirculating systems. Recirculating systems are also used in the hatchery and larval-rearing stages for several aquaculture species. Overall, about 2% of United States aquaculture production by weight derives from recirculating systems, representing about 4% of production by value.

Open-water molluscan shellfish culture

Clams, oysters, mussels, and other molluscan shellfish are cultured in coastal embayments and estuaries along the Atlantic, Gulf, and Pacific coasts of the United States. Open-water mollusk culture is unique among other types of United States aquaculture because mollusks obtain their food by filter-feeding phytoplankton and particulate matter from the water, rather than being fed manufactured feeds. Although there may be some artificial feeding of larvae or juveniles in hatcheries, no feed costs are incurred during the grow-out phase and, on a net basis, mollusk culture removes nutrients and organic matter from surrounding waters.

Most oysters and clams are grown in on-bottom aquaculture, which involves seeding mollusks into prepared grow-out areas where they are allowed to mature to harvestable size. Off-bottom culture involves suspending mollusks in water column cages, racks, or bags, or attached to stakes or ropes. Off-bottom culture is not as common in the United States as it is in other parts of the world. Regardless of the method, cultured mollusks are grown in waters leased from the state; privatization through leasing provides the farmer with control over culture activities and protection of stock from poaching.

Open-water molluscan shellfish culture accounts for about 25% of United States aqua-culture production by weight and 19% by value.

AQUACULTURE AND THE ENVIRONMENT

The Global Context: Human Dominance of Environmental Change

Human activities now dominate land transformations, alterations in global biogeochemical cycling of water, carbon and nutrients, and extinctions and invasions of species (Vitousek et al. 1997). Humans appropriate most of the renewable supplies of freshwater (Postel et al. 1996), from 10 to 55% of terrestrial net primary production (Rojstaczer et al. 2001) and 24 to 35% of the net primary production from freshwater, continental shelves, and upwelling areas (Pauly and Christensen 1995). The most serious global environmental problems—climate change and loss of biodiversity—are in large measure caused by human activities. The increased demand for ecosystem services that supply food, freshwater, timber, and hydropower has been met by humans consuming a greater fraction of the available supply of these services and increasing the production of others through new technologies and expanded area, as in the case of food produced by aquaculture. The Millenium Ecosystem Assessment (MEA 2005) focused attention on the relation-ship between human well-being and ecosystem services. Ecosystem services are the sup-porting, provisioning, regulating, and cultural benefits that humans obtain from ecosystems. Supporting services, such as photosynthesis, primary production, nutrient cycling, water cycling, and soil formation, are fundamental because they affect the ability of ecosystems to provide other kinds of services. Provisioning services are the products from ecosystems, such as food, freshwater, wood products, fiber, and genetic resources. Regulating services include climate, flood, water purification, waste treatment, and the effects of disease and pests. Cultural services include educational, recreational, aesthetic, and spiritual services.

Collectively, these services support human well-being, including the basic materials for living, personal health, security, and good social relations.

Changes to ecosystem services are affected by indirect and direct drivers. Classes of indirect drivers include demographic, economic, sociopolitical, science and technology, and cultural and religious beliefs. These in turn affect direct drivers of changes to ecosys-tem services. Examples of direct drivers include changes in land use, species introductions, applications of external inputs, resource consumption and harvest, use of particular tech-nologies, climate change, and natural disasters. Two of these drivers—nutrient loading and global climate change—are expected to increase in importance in the future. The interactions among indirect drivers, direct drivers, ecosystem services, and human well-being occur across local, regional, and global spatial scales and across short- to long-term temporal scales.

For terrestrial ecosystems, the major direct driver of ecosystem change has been the conversion to cropland and the application of improved technologies that have increased the supply of food. For marine ecosystems, the major direct driver has been fishing, espe-cially in coastal areas (about half of marine stocks are fully exploited). For freshwater ecosystems, the most important direct drivers have been flow modification of running waters, invasive species, and nutrient pollution. Coastal ecosystems are affected by eutrophication; habitat loss; invasive species; and pollution from adjacent land, upstream sources, or the marine environment. Nutrient loading in excess of carrying capacity has emerged as one of the most important direct drivers of change in terrestrial, freshwater, and coastal ecosystems. On the one hand, increased nutrient applications have resulted in increased food production, but the eutrophication that results from excess nutrients enter-ing freshwater ecosystems has taxed the water purification ecosystem service. Increased nutrient loading has resulted in the greater frequency of harmful algae blooms, hypoxic zones, fish kills, human health problems, and damage to light-sensitive coastal ecosystems such as seagrass meadows and coral reefs. It is important to note that aquaculture can be a contributor to increased pressures on each of these ecosystems, but it can also be affected by a driver in a general way.

The MEA (2005) found that 60% of ecosystem services examined (15 of 24) are being degraded or used unsustainably, including freshwater supply, capture fisheries, waste treat-ment, and water purification. Capture fisheries and freshwater are now used at levels that exceed supply. Maximum global fish catch occurred in the late 1980s and is now declin-ing. The mean tropic level of the fisheries catch is also declining (Pauly et al. 1998). Currently, more than 25% of fish stocks are overexploited or depleted. Much of the deg-radation of ecosystem services is a result of mankind's exploitation of ecosystems to meet the increased demand for food, freshwater, energy, and other natural resources. Although 60% of ecosystem services have been degraded, 4 of 24 have been enhanced, including food production from agriculture and aquaculture.

The Sectoral Context: The Environmental Impacts of Aquaculture

Concern over the environmental impacts of aquaculture is a relatively recent phenomenon. Aquaculture has a long history, particularly in Asia, and during most of this time, the environmental effects of aquaculture production were perceived to be acceptable. It was not until global aquaculture production increased rapidly in the 1980s and 1990s that local aquaculture industries grew sufficiently large to attract the attention of environmentalists, regulators, and consumers. In particular, the rapid growth of penaeid shrimp farming in tropical coastal ponds and salmon farming in marine net pens has stimulated concern about the environmental sustainability of these forms of aquaculture, and by extension, all of aquaculture (Pillay 2004).

The environmental effects of aquaculture are value-based, which is to say that the positive and negative effects of aquaculture are determined by societal values as shaped by factual knowledge and individual perceptions. Aquaculture has known benefits to society, such as the increased availability of high-quality animal protein, poverty allevia-tion, increased employment, foreign exchange earnings for developing countries, and profit for entrepreneurs and investors. Aquaculture also has societal costs, as exemplified in conflicts over access to resources and land traditionally considered to be common property or open access. Certain forms of aquaculture are also seen to be environ-mentally benign, or even beneficial, while others are not. For example, the aquaculture of shellfish and seaweeds can mitigate the eutrophication of coastal waters. However, most forms of aquaculture are perceived to have some adverse environmental effects.

Potential environmental effects depend on the trophic level of the cultured species, culture system type, production intensity, and extent or concentration of landscape devel-opment in aquaculture. In general, culture of species at higher trophic levels, culture systems that are more open to the environment, intensive culture systems, and high con-centrations of aquaculture facilities are seen to have adverse impacts on the environment. The deleterious environmental effects of aquaculture on the aquatic environment should be seen in the context of other assaults. The intensification of agriculture (especially animal agriculture), deforestation, industrialization, and urbanization contribute to the degradation of aquatic environments. Thus, many environmental impacts have multiple causes, of which aquaculture is one possible contributor.

Environmental impacts of aquaculture can be divided into near-field and far-field effects. Near-field effects are the localized effects of aquaculture production, which are, in most cases, reversible. Examples of near-field effects are benthic disturbances beneath net-pen facilities and habitat conversion when ponds are constructed. Near-field effects have been the best-studied, primarily because they are more amenable to evaluation. Far-field effects of aquaculture are less well understood, primarily because they are often one of many sources of impact and estimating the relative contribution of aquaculture to environmental change is difficult. Examples of far-field effects are introduction of nonnative species, pathogen exchange between captive and wild populations, and effects of predator control at aquaculture facilities on wildlife populations and biodiversity. The uncertainties associated with far-field environmental effects of aquaculture are large.

Improving aquaculture's environmental performance requires solutions implemented at spatial scales ranging from individual culture units to the entire biosphere. Solutions also range from farm-level implementation of improved practices to fundamental changes in social and economic values. This conceptual distinction—where solutions to environmen-tal problems may be characterized as either technological or value-oriented—corresponds to broader discussions of the relative roles of technocentric and ecocentric approaches to achieving sustainability of human activities (O'Riordan 1981). These themes have also been compared in the context of economic growth, where they help define contrasting views of the economics of resource depletion and the value of technology and "man-made capital" in offsetting resource degradation (Hediger 1999). Recently the concepts of eco-centrism and technocentrism have been used to help defi ne and contrast the concepts of "sustainability" and "sustainable development" (Mebratu 1998; Robinson 2004).

Ecocentric approaches to sustainability are loosely defined by ecosystem preservation, changes in social values and lifestyles, and minimizing resource use. Issues with ecocentric-dominated solutions tend to have large spatial scales and are future-oriented. Global climate change is a familiar example of an issue requiring ecocentric-dominated solutions. Addressing climate change requires fundamental changes in social and economic values, and a highly developed concern for the well-being of future generations. Technocentricism is characterized by ecosystem conservation, technology development, and improving efficiency of resource use. Technocentric solutions to environmental prob-lems also tend to focus on issues of smaller spatial scale and of more immediate concern. Although these distinctions have been widely discussed, refined, and debated—especially in the literature of ecological and social economics—it should be clear that they are artificial typologies and that the full range of environmental impacts from most human activities—including agriculture and aquaculture—can be addressed only in the context of both philosophies. For example, global climate change, while clearly a future-oriented problem requiring fundamental changes in social and economic values to address, will nevertheless rely on technology to provide, as one of many examples, alternative energy sources with lower greenhouse gas emissions.

As such, ecocentrism and technocentrism are merely viewpoints rather than alternative solutions, both of which must be integrated into a broad approach if the impacts of human activities are to be reduced so that we preserve capacity to provide for the well-being of future generations. Distinctions between ecocentric and technocentric solutions to envi-ronmental problems (and, in a broader context, as means to achieve sustainability) do, however, serve as useful counterpoints to arrange the remainder of this chapter. The fol-lowing section describes impacts related to broad-scale resource use. Following that, we describe impacts that are more closely related to facility operations. Both are amenable to the improvement of the environmental performance of aquaculture through better manage-ment practices.

Addressing the basic problems related to resource use ultimately depends on shifts in social and economic values, particularly for people in the developed nations of the world. Using salmon net-pen aquaculture as an example, energy efficiency can be improved at the farm level by adopting improved technologies, such as more effi cient equipment and husbandry practices that improve feed utilization, growth rate, and survival of fish. These are technocentric solutions and are important in conserving resources as well as improving the economic performance of the facility. However, the reduction in energy use (per unit protein production) achievable by adopting improved technologies is at least an order of magnitude less than the reduction achievable by farming animals from lower trophic levels, such as tilapias or carp.

Choosing to farm carnivorous aquatic species, such as salmon, is based on consumer demand and willingness to pay the relatively high marketplace price that is, in part, related to the high energy costs of production. As long as consumers are willing to pay for aqua-culture products with relatively high resource-input requirements, someone, somewhere, will produce them. Radical changes in overall energy use in aquaculture will occur only when social values change (that is, consumers demand "low trophic-level fish" based on their concern for the environment) or when changing economic conditions drive up the cost of production to the point where species with high energy-input requirements become too expensive to purchase. An analogous situation exists in the argument for large-scale adoption of vegan lifestyles based on resource use and environmental impacts associated with food production. Although protein choice significantly influences agriculture's envi-ronmental performance (Goodland 1995; Goodland and Pimentel 2000; Reijnders and Soret 2003), lifestyle changes on a scale great enough to have significant impact on global energy use in agriculture are difficult to envision until induced by market pressures.

RESOURCE USE IN AQUACULTURE

The major physical resources required in aquaculture are energy, land, water, and feed. All are finite and the impacts of aquaculture on resource availability depend on the overall rate and efficiency of use. Aquaculture may also impact resource availability by altering, rather than using, a resource. For example, flow-through aquaculture systems consume almost no water but they alter its quality by adding wastes produced during culture. Effects of resource alteration must be managed so aquaculture does not diminish the value of the resource for some other use.

Resources are also subject to conflicting demands on their use and, ultimately, these conflicts are difficult to address because they involve trade-offs. The fundamental question is whether the use of a particular resource in aquaculture has more social or economic value than its use in some other activity. The conflicting or alternative use might be in another food-producing sector of agriculture or it might be an environmental function, such as the biodiversity afforded by a wetland that is being considered as a site for an aquaculture facility. Some conflicts can be addressed through technology or improved management, but many are resolved only through economic incentives and market mech-anisms, changes in how society values a resource, or development of environmental policies or regulations by governments or other institutions.

Energy

Aquaculture, as is true of agriculture in general, is a process whereby solar and fossil fuel energy are transferred from the environment into the culture system and converted into food energy. Although this concept oversimplifies the process because it ignores nonenergy aspects of food quality, it is conceptually useful because most inputs to food production can be expressed in energy units, thereby making it possible to compare resource use among various aquaculture and agriculture production systems. For example, water use can be converted to energy units by quantifying the energy used to pump water, as well as the energy required to build and install pumps, pipes, and other infrastructure related to the water supply. Quantifying energy use across various aquaculture production activities (production of juveniles, facility construction, feeds and feeding, labor, process-ing, and so on) also provides a means of identifying production inefficiencies where alternative practices may lead to energy and costs savings.

The two primary energy sources for aquaculture are solar radiation and fossil fuels. No aquaculture system relies entirely on solar radiation, although sunlight is ultimately the major source of energy for all agriculture because plant material produced in photosyn-thesis is the base for nearly all food chains. But even in extensive cultures of seaweed and other aquatic plants where solar energy fully supports crop growth, fossil fuels are expended in functions such as harvesting and processing. Likewise, certain types of shellfish aqua-culture can be highly energy effi cient because fi lter-feeders grow by consuming phyto-plankton swept past the beds or farms by tidal currents, but fossil fuel expenditures are made in the hatchery, harvesting, and processing phases of production and human labor is required during production, harvesting, and processing.

Traditional pond aquaculture, as originally practiced in China, relied heavily on solar energy incident on the culture system to provide food, via photosynthesis, to support fish production. Primary production was enhanced by fertilizing the system with plant nutrients derived from terrestrial plants, animal manures, or processing by-products of agricultural crops. Two or more fish species, usually carp, were cultured together (*polyculture*) to make effi cient use of the variety of natural foods produced within the pond. As originally con-ceived, fish were locally consumed as fresh product, which reduced energy costs for transportation and eliminated energy costs for storage.

In the last half of the 20th century, global aquaculture transitioned from low-input systems relying heavily on solar radiation incident on the culture facility to systems requir-ing greater imports of energy from other sources. The goal of aquaculture shifted from low-intensity subsistence production of diverse species to intensive aquaculture focused on maximizing economic benefits through increased yields. As the goals of aquaculture changed, aquaculture systems increasingly relied on fossil fuels and appropriation of fixed solar energy from other ecosystems.

Energy inputs

Energy used in fisheries and aquaculture can be categorized as 1) ecosystem support; 2) direct energy inputs, and 3) indirect or embodied energy (Troell et al. 2004; Tyedmers 2004). *Ecosystem support* accounts for solar energy fixed in other ecosystems and then "imported" to the aquaculture system. An example of ecosystem support is the solar energy used in photosynthesis to produce soybeans and other plant feedstuffs that are used in manufactured feeds. Similarly, photosynthesis by marine phytoplankton is the base of the food web culminating in pelagic marine fish harvested for fishmeal and fish oil. Ecosystem support also includes energy used by natural processes to assimilate wastes produced during culture. For example, waste nitrogen and phosphorus produced during aquaculture may be discharged to a wetland where they are assimilated by plants. A certain amount of solar radiation and ecosystem area is needed to support the plant growth required to assimilate those nutrients. In general, aquaculture systems relying on manufactured feeds to support production (which includes much of finfish aquaculture in the United States) and those systems that discharge directly to public waters (net pens and flow-through systems) require a large ecosystem area outside the culture system to concentrate solar radiation into the chemical energy of feedstuffs and to process and assimilate wastes pro-duced during culture.

Direct energy inputs are predominantly fossil fuels and labor used to provide most other resources needed in production. Examples include easily measured farm-level inputs such as energy used for pumping water and mechanical aeration. Other direct energy inputs are not so obvious. For example, a significant input to many types of aquaculture is fuel and labor used by farms and mills to produce and then process plant feedstuffs into aquafeeds. Another important set of energy inputs is the fuel, electrical energy, and labor used by fishing fleets and fish reduction plants to provide fishmeal. Direct energy inputs are highest for culture systems that rely on high-quality aquafeeds and in recirculating aquaculture systems where waste processing requires fossil fuel input.

Indirect energy inputs, also called *embodied energy inputs,* account for the energy needed to construct and maintain tanks, nets, ponds, buildings, pumps, aerators, boats, and other fixed assets needed to produce the aquaculture crop. Depending on the extent to which energy inputs are accounted, indirect energy costs may even include the energy and labor needed to obtain the raw materials—such as steel, wood, and plastic—used to fab-ricate the asset. Indirect energy inputs are highest for intensive aquaculture facilities that have significant physical infrastructure, but indirect energy inputs typically represent only a small fraction of overall energy costs of intensive aquaculture production because of the overwhelming energy inputs associated with production of manufactured feeds.

Life-cycle energy assessment

Life-cycle assessment (LCA) is a scientific discipline that attempts to account for all environmental impacts of providing specific goods and services to society (Pennington et al. 2004; Rebitzer et al. 2004). Life-cycle assessment derives its name from the concept that all products have a "life" starting and ending at predefi ned points that set boundaries for the assessment. Product life in agriculture may, for example, begin with acquisition of raw materials (fertilizers, feedstuffs, etc.) and include production, processing, transporta-tion, and so on

Assessments can be made on the basis of any impact of interest. For example, total contribution to greenhouse gases can be estimated for all activities involved in production and retirement of a particular product. Other commonly used impacts in LCA include eutrophication, ozone creation, ecotoxicology, carcinogen production, land use, and water use. One of the most useful, and common, bases for LCA is energy input over product life. Energy as a basis for LCA is appealing because it is the simplest and most intuitive "common currency" for comparing impacts of all activities involved in production, use, and retirement among various products, processes, or activities.

Studies using LCA methodology consistently show that the energy used to produce manufactured feeds dominates the energetics of many modern aquaculture production systems. For example, more than 75% of the total energy cost of producing Atlantic salmon in net pens is used in procuring or growing feed ingredients and manufacturing the feed (Folke 1988; Troell et al. 2004; Tyedmers 2004; Ellingsen and Aanondsen 2006). The remaining energy inputs, in order of importance, were fuel and electricity used to operate the facility, embodied energy costs (manufacture, maintenance, etc.) associated with physical infrastructure, and energy used to produce smolts. Feed production domi-nates energy budgets of all aquaculture systems relying on manufactured feeds, regardless of species, and overall energy use per unit protein production decreases in aquaculture systems less reliant on manufactured feeds (Troell et al. 2004).

Energy use in aquaculture

The energy efficiencies of various aquaculture systems span perhaps the widest range of any agricultural sector (Fig. 1.4) (Troell et al. 2004; Tyedmers 2004). Traditional carp polyculture in ponds fertilized with agricultural by-products lies at one end of the spectrum as one of the most energy-efficient food production systems ever devised. When energy efficiency is expressed on the basis of industrial energy input per unit of edible protein energy produced, traditional carp polyculture rivals even vegetable crops for energy efficiency, with energy input/output ratios approaching 1. Most agricultural activities have energy input/output ratios much greater than 1, meaning that energy inputs during production far exceed food protein energy output. Relative to traditional pond culture of carp, aquaculture systems that rely heavily on manufactured feeds and other ecosystem support functions lie at the other end of the spectrum, with energy input/ energy output values exceeding 50. Nonetheless, most modern aquaculture systems are generally com-parable to terrestrial animal production systems with respect to energy efficiency. For example, input/output energy ratios for pond-grown tilapia and channel catfish are similar to those for several common animal production activities, such as eggs, poultry (broiler), and swine production (Troell et al. 2004; Tyedmers 2004). Likewise, input/output energy ratios for marine shrimp aquaculture, which are among the highest of the major aquacul-ture systems, are in the same general range as that for shrimp trawling.

Comparing energy use in aquaculture and other forms of agriculture is difficult—and sometimes misleading—because few energy analyses of aquaculture have been conducted. Further, studies may not include adequate information on the boundaries of the analysis, which will bias comparisons among systems. Depending on the intent of the analysis, an almost endless number of input functions can be included in an LCA for an agricultural activity. Most studies of energy use include only the most easily quantifi ed inputs, such as direct electrical and fuel inputs for the most obvious production functions (such as feed manufacture, water pumping, aeration, and so on).

Life-cycle assessment of energy use can include postharvest functions such as process-ing, freezing, refrigeration, storage, transportation, marketing, waste treatment, and even household activities such as refrigeration, freezing, and cooking. Energy use in these activities apparently has not been assessed for aquaculture but may be an important part of the overall energy costs of delivering aquaculture products to a consumer's plate. For example, energy used in on-farm production of the United States food supply accounts for only about 20% of the energy used to deliver food to the consumer's plate (Heller and Keoleian 2000). Postharvest processing and transportation each consume about 15% and household preparation accounts for more than 30% of the total energy consumed. Ultimately it will be economically and socially imperative to improve the energy efficiency of all aspects of the food-supply chain. However, it is possible that greater overall gains in energy savings can be made by improving the effi ciencies of processing, transport, retail-ing, and even household storage and preparation than can be made by improving energy efficiency in the production sector. This may have particular relevance to aquaculture, where important products are produced only in certain regions (marine shrimp in the tropics; salmon in the north-temperate) and are stored and shipped long distances for consumption.

Land

Aquaculture uses land in two ways. First, aquaculture facilities occupy a defi ned area or space on land or in water. But facility area accounts for only a portion of the total land or water area needed to produce an aquaculture crop. Additional ecosystem area is needed to provide support or service functions. The two most important of those functions are food production and waste treatment.

Facility area

The area occupied by aquaculture facilities in the United States varies widely. There are approximately 125,000 ha of commercial aquaculture ponds in the United States, with 50,000 ha devoted to channel catfish culture. Contiguous blocks of catfish ponds in Mississippi may cover several hundred hectares on individual farms. By comparison, the total United States production of Atlantic salmon is derived from only a few hundred hectares of net cages on the Atlantic and Pacifi c coasts.

The land or water area needed to produce a unit of aquaculture crop is inversely pro-portional to production intensity. Production intensity is a vague term that attempts to describe the nature of resource use in aquaculture production. A simple, and common, measure of intensity is crop yield per unit of resource use, such as land. The land or water area needed per unit production of aquaculture crop varies over more than two orders of magnitude. At one extreme are highly intensive water recirculating aquaculture systems (Chapter 10), which are capable of annually producing 1 to 2 million kg of fish per hectare of culture unit (Timmons et al. 2002). Rainbow trout production in raceways (Chapter 9) is about 10 times less intensive and therefore requires about 10 times the surface area per unit of fish production. Fish and shrimp production in ponds (Chapters 6 and 7) requires several hundred times the land area compared with intensive recirculating systems.

Differences in the relative sizes of aquaculture facilities (or, more simply, the intensity of the system) are not simply a function of one system being inherently more effi cient than another. Rather, the area required to produce a unit crop yield depends on the extent to which food production and waste treatment are subsidized by ecosystems or processes external to the culture system. For example, much of the ecosystem support for traditional pond aquaculture is inherent in the system. Traditional aquaculture ponds used in Chinese carp polyculture function not only to confine fish, but also to provide an internal area for food production and waste treatment. Land requirements for pond aquaculture are therefore relatively large. In net-pen culture, on the other hand, the culture unit functions only to confine the crop. Food is produced and wastes are treated in separate, external ecosystems. As such, the relative area needed for a net-pen facility is much smaller than for ponds.

Ecosystem support: Food

Many, if not most, aquaculture systems in the world rely on plant photosynthesis either within the culture unit (ponds) or in nearby waters (open-water molluscan shellfish culture) to produce most of the food that supports aquaculture production. In ponds relying on *autochthonous* (within-pond) food production, aquaculture yield is limited by the rate of primary production, which in turn is ultimately limited by the amount of solar radiation impinging directly on the culture unit. Such systems are usually fertilized with plant nutri-ents to make fullest use of incident solar radiation and often contain two or more her-bivorous and omnivorous species to make effi cient use of the wide variety of natural foods produced in fertilized ponds. Production in molluscan shellfish aquaculture depends on solar radiation over a wide area to produce natural foods that are swept past and captured by the fi lter-feeding organisms.

Production of herbivorous or omnivorous fish can be quite impressive (3,000 to 10,000 kg/ha per year, or more) in fertilized ponds. The food web in these systems is relatively simple, and solar energy captured by plants is transferred effi ciently to fish at these lower trophic levels. To increase aquaculture productivity past that achievable in systems dependent only on autochthonous primary production, food produced outside the culture system must be imported. In some systems, that food may be low-quality organic matter that might otherwise be considered a waste product, such as agricultural by-products or livestock manures. In many aquaculture systems worldwide—and in nearly all commercial systems in the United States—the *allochthonous* (from outside the system) organic matter added to increase per-area production consists of high-quality manufactured feeds made from plant (soybean and corn, for example) and animal (usually fish) meals. Production of feedstuffs requires land external to the culture unit.

Soybean meal, wheat middlings, cornmeal, and cottonseed meal are common plant products used in aquafeeds. Boyd et al. (in press) calculated land areas needed to procure terrestrial feed ingredients to produce 1 tonne of various aquaculture species by using typical aquaculture production data, feed composition, and information from the United States Department of Agriculture (USDA 1994) for average plant seed and meal yields. Channel catfish is an example of a species grown on feeds with high levels of plant mate-rials (>85% of the diet) and low levels of fishmeal and fish oil (<5% of the diet). Producing 1 tonne of channel catfish requires about 0.4 ha of cropland to produce plant feedstuffs. Atlantic salmon are fed diets with relatively low amounts of plant ingredients (20 to 40%) and high amounts of fishmeal and fish oil (60 to 80% of the diet). Atlantic salmon produc-tion requires a land area of about 0.1 ha/tonne for plant feed ingredients.

Marine ecosystems are required to produce small pelagic fish for reduction to meal and oil for inclusion in manufactured feeds. The appropriated marine ecosystem area can be calculated from the primary productivity of the fishing area and the trophic level of the pelagic fish captured for reduction. Values range from less than 10 to more than 100 ha of marine ecosystem area per tonne of fishmeal produced (Folke 1988; Larsson et al. 1994; Tyedmers 2000). Assuming a value of 50 ha of marine area per tonne of fishmeal, and typical feeds and production conditions used by Boyd et al. (in press), approximately

2.2 ha of ocean area is needed to provide fishmeal to grow 1 tonne of channel catfish and about 18 ha of ocean is needed to grow 1 tonne of salmon. These values are indicative rather than defi nitive because marine areas needed for fishmeal production vary greatly depending on the type of fishmeal used, the productivity of the fishery, and feed conver-sion when fed to fish. For example, the value of 18 ha/tonne calculated here for net-pen cultured Atlantic salmon is higher than the value of 14.2 ha/tonne reported by Tyedmers (2000) and much lower than the value of 100 ha/tonne reported by Folke (1988). The qualitative implications of this exercise are 1) providing terrestrial ingredients for fish feeds appropriates relatively small areas of land whereas providing fish-derived feedstuffs requires relatively larger ecosystem support areas and 2) total appropriated ecosystem area for food production in intensive aquaculture is reduced when animals, such as channel catfish, from lower trophic levels are cultured.

Ecosystem support is often expressed on the basis of hectare of support area per hectare of culture area. When expressed in this manner, the dependence on ecosystem support varies over four orders of magnitude, depending upon culture system and feed composi-tion. Intensive net-pen culture of carnivorous fish, such as salmon, requires an ecosystem support area for food production that is more than 10,000 times the area of the culture system (Folke 1988; Folke et al. 1998). Growing the omnivorous species, channel catfish, in ponds appropriates an area for procurement of feedstuffs that is approximately 10 times the area of the pond (recalculated from Boyd et al., in press, using additional data on ecosystem areas needed for fishmeal). At the extreme, carp, tilapia, and other fish from lower trophic levels can be grown in ponds fertilized with agricultural wastes or by-products, and do not require external ecosystem area for food production (Berg et al. 1996). Those systems have a value of 1.0 for the ratio of ecosystem support to culture area.

Open-water molluscan shellfish culture is unique in that animal growth depends on natural foods produced in marine or brackishwater ecosystems that are much larger than the culture area. Those foods are swept over the culture area by tidal currents and removed by the fi lter-feeding animals. In one context, mollusk culture "appropriates" food produced in an area much larger than the culture area, but this food is not cultivated and requires no additional input of resources past solar radiation and nutrients already present in the water. Natural food organisms consumed by mollusks and the nutrients that supported their growth are usually considered underutilized, and their removal from the ecosystem is seen as beneficial. Accordingly, no additional ecosystem support area should be assigned to the fi nal grow-out phase of open-water mollusk culture.

Ecosystem support: Waste treatment

With the exception of molluscan shellfish culture, all aquaculture systems use resources procured from one ecosystem, concentrated, and then added to the aquaculture system, inevitably producing waste. In addition to land area for facilities and to produce food, ecosystem area is therefore required to assimilate those wastes. In recirculating aquaculture systems and ponds operated with long hydraulic residence times, significant quantities of waste produced during culture are treated within the facility and there is relatively little external area needed for waste treatment. On the other hand, much of the waste produced in raceway and net-pen culture is discharged directly to the outside environment. The ability of the external ecosystem to assimilate those wastes may limit aquaculture produc-tion either by polluting the surrounding water to the point where animal welfare inside the facility is endangered (*self-pollution*) or by imposing regulatory constraints on the amount of waste that can be discharged. In addition to effects on aquaculture production, waste discharge into public waters may create societal externalities such as degraded water quality that limit options for use, water treatment costs, and other downstream impacts.

The ecosystem area needed for waste assimilation, expressed as hectares of waste treat-ment area per hectare of production facility, varies over at least two orders of magnitude, depending on the type of production system. At one extreme are aquaculture ponds with low to moderate stocking and feeding rates that can, in theory, be operated for many years without intentional water exchange to remove wastes. Natural biological, chemical, and physical processes active inside the pond remove or transform wastes at rates adequate to prevent year-to-year accumulation of potential pollutants. In theory, no outside ecosystem support area is needed to treat wastes produced during culture and, therefore, the ratio of land area for waste treatment divided by facility area is one. In other words, the pond functions as its own waste treatment facility. In practice, of course, ponds must be occa-sionally drained and some overflow is inevitable during periods of heavy precipitation. Nevertheless, for ponds operated with long hydraulic residence times, more than 90% of the waste organic matter, nitrogen, and phosphorus produced during culture is assimilated inside the pond before water is discharged (Tucker et al. 1996).

Internalizing waste treatment imposes a relatively high direct land cost to pond aqua-culture, and this cost is evident in the large size of pond facilities (Fig 1.5). Extensive land use is the result of the pond functioning as both an animal confinement area as well as a waste treatment facility. For example, more than 95% of the total area of a catfish pond functionally acts as a photosynthetic waste-treatment lagoon and less than 5% of the total area serves as a "fish-holding" area (Brune et al. 2003). In effect, more than 95% of the land and construction costs for ponds can be assigned to waste treatment functions, and the relatively large land area occupied by a pond aquaculture facility is a price the farmer pays for treating wastes on site rather than discharging the waste to public waters. Engle and Valdarrama (2002) explored other costs (such as aeration, labor, etc.) associated with internalizing waste treatment in channel catfish ponds and calculated that almost 30% of the total cost of producing channel catfish can be ascribed to internal waste treatment processes. Land area requirements and other costs resulting from internalizing waste treat-ment in ponds limit profitable culture only to areas where large tracts of fl at land are available at a reasonable price.

Aquaculture production in ponds is therefore limited by the finite capacity of the pond ecosystem to treat wastes produced during culture (Hargreaves and Tucker 2003). Further intensification of production is possible only if wastes are treated external to the culture unit, usually by discharging wastes to public waters. Accordingly, ecosystem area outside the culture facility is used to assimilate wastes, and in most instances the cost of that treatment is borne by society, not the aquaculturist.

Relatively few studies have been conducted to determine the external ecosystem areas needed to treat aquaculture wastes produced by aquaculture in raceways or net pens. Furthermore, requirements will vary depending on the hydrology and biology of the eco-system into which wastes are discharged. Based on the few studies conducted, it appears that ecosystem support areas between 100 and 300 times the facility area are needed to treat wastes produced from cage and net-pen fish culture (Berg et al. 1996; Kautsky et al. 1997; Folke et al. 1998; Brummett 1999). These values for ecosystem waste assimilation area are somewhat larger, but still within the same approximate order of magnitude as the ratio of waste treatment area to fish-holding area in the catfish ponds described above. This is not a coincidence, because the same biological and physicochemical processes are responsible for waste treatment whether the water is inside a pond or in the lake, stream, or estuary in which cages are suspended. Areal requirements for waste treatment should therefore be of similar magnitude. The difference, of course, is that the ecosystem area needed for waste treatment in ponds is, for the most part, inherent in the production system, whereas waste treatment for cage and net-pen culture is external to the system.

Mollusks generate wastes in the form of feces and pseudofeces that are deposited in sediments around shellfish culture areas (see Chapter 11). Those wastes may impact the local environment and ecosystem area is required to assimilate those wastes. However, on a larger scale, mollusks are net consumers of particulates and nutrients, so in a sense, the local ecosystem area needed to assimilate wastes produced in the shellfish-growing area is less than the overall ecosystem area that would have been required to assimilate the waste in the absence of the fi lter-feeding shellfish. The concept of ecosystem support required to treat wastes is difficult to apply to open-water molluscan shellfish aquaculture.

Land-use issues in aquaculture

Regions suitable for land-based aquaculture are limited by the right combination of climate, soil type, geomorphology, water availability, and access to other resources such as labor, markets, feeds, and larvae or juveniles for stocking. Nevertheless, land use per se is seldom a limiting factor in aquaculture development. This is particularly true in the United States, where relatively few locations are significantly developed into aquaculture. The Yazoo-Mississippi River alluvial valley in northwest Mississippi is the most intensely developed aquaculture region in the United States, with aquaculture ponds occupying more than 45,000 ha of land. Yet ponds account for only about 2% of land use in the region. Even in sub-watersheds within the Yazoo basin with more intense development, ponds account for only 10 to 20% of the watershed (Fig. 1.6). Aquaculture development in the United States is more likely constrained by water availability, regulatory constraints on land development, or the value of land for alternative uses than by the availability of suitable land for facility construction.

At times, availability of the right combination of resources leads to localized overde-velopment of aquaculture. In this respect the problem is not necessarily the physical concentration of facilities, but rather ecological imbalances associated with the inability of the local ecosystem to provide support services, such as adequate water supply or waste treatment. A good example is the study of the ecosystem support area needed for semi-intensive shrimp farming in the Bay of Barbacoas area of Colombia, South America (Larsson et al. 1994; Kautsky et al. 1997). The estimated local ecosystem support area needed to assimilate waste nutrients from shrimp farms closely matched the mangrove area available to provide that service. Local capacity for waste treatment was used to its fullest extent and additional development could lead to self-pollution of the water supply, spread of disease, and other environmental impacts that have been associated with the collapse of shrimp farming in other overdeveloped areas (Primavera 1993).

Most problems associated with land use in aquaculture are related to the value of alter-native uses of the land. These conflicts arise when aquaculture must compete with other land uses. Land-use conflicts tend to be especially acute in coastal regions, which are limited in area, have proportionately large areas of sensitive and environmentally valuable ecosystems, and are intensively used for a range of often conflicting purposes by dense human populations. Coastal regions also tend to be more commercially developed and the land more costly, especially in the United States.

Clearing mangrove forests to build ponds for shrimp and fish culture has been one of the most contentious land use conflicts in aquaculture (Primavera 1993; Naylor et al. 1998; Boyd 2002). Mangrove ecosystems have high inherent aesthetic and environmental value. They protect coastlines from erosion and fl ooding and provide habitat and food for a tremendous variety of aquatic, terrestrial, and avian species. Mangroves also act as natural purification systems for waters interacting at the interface of freshwater and marine envi-ronments. They are also convenient—although far from ideal—locations for brackishwa-ter aquaculture ponds, especially for growing shrimp. In the 1980s and early 1990s, when global shrimp aquaculture was expanding rapidly, thousands of square kilometers of ponds were built in mangrove areas. Much of this land had been cleared previously for other uses, but the destruction was often attributed to shrimp farming. Although environmental advocacy groups overstated the involvement of shrimp farming in the historic loss of mangrove, worldwide awareness of the problem resulted in stricter governmental oversight of the use of mangrove forests and stimulated significant changes in shrimp production practices to protect these valuable ecosystems (Boyd 2002; FAO 2006b).

Aquaculture can affect sensitive ecosystems other than mangroves. Marine benthic communities may be impacted by net-pen aquaculture, and construction and operation of freshwater ponds can destroy sensitive wetlands. Certain types of open-water molluscan shellfish culture—usually viewed as an environmentally benign endeavor—can damage seagrass beds due to siltation and the mechanical effects of certain culture practices.

Marine aquaculture also faces unique conflicts unrelated to the physical area occupied by the facility or the impacts of culture activities. Confl icts may arise from competition with other users of the area or region, including activities such as tourism, shipping and navigation, capture fisheries, and urban development. Confl icts may even arise from esthetic considerations, including the effect of aquaculture development on the scenic quality of an area (the *viewscape*) and disturbances related to increased activity and noise associated with culture practices.

Nearly all problems arising from land-use conflicts in aquaculture can be avoided by prudent site selection. Better management practices for facility site selection are described in each of the production system chapters of this book (Chapters 6 through 11).

Water

Agriculture and water resources

Cultivated systems, including freshwater aquaculture, now cover 24% of the terrestrial surface of the planet. Most of the increase in food production in the last 50 years has been derived from agricultural intensification, not an increase in cropland area. Intensifi cation has increased pressure on water resources (both for supply and waste treatment) and fisheries and cropland resources for feed, and it requires increased energy resources, nearly all derived from fossil fuels.

Freshwater is a scarce resource on the planet, with less than 1% of total freshwater resources available for human use. Renewable freshwater resources available for human use are derived from precipitation that falls on land. This precipitation runs off (including infi ltration to groundwater) or is lost by evapotranspiration from land and water surfaces and through plants. Humans use about 10% of renewable freshwater supplies (Postel et al. 1996), although some countries use more than 100% of renewable supplies. About 40 to 50% of freshwater runoff is appropriated for human use. The available supplies of global freshwater are used by agriculture (75%), by industry (20%), and for domestic purposes (5%) (UNEP 2002).

Aquatic resources are impacted by numerous human activities. Running waters are exploited to supply irrigation and drinking water, generate electricity, and receive wastes. Running waters are also affected by land use in adjacent watersheds. Threats to running waters include alteration of habitat; changes in water chemistry; and changes in biodiver-sity caused by land transformations, channelization, and industrial or urban development. Standing waters provide water for agricultural, industrial, and domestic uses. Threats to standing water ecosystems include eutrophication, chemical contamination, and the intro-duction of invasive species.

Water use in aquaculture

Water is an obvious and essential resource for aquaculture. No single factor influences the success of aquaculture more than availability of an adequate supply of good quality water. Availability of freshwater increasingly constrains all forms of human development, so aquaculture must compete with other uses, including irrigated agriculture, industry, and urbanization. Competition for finite water supplies is particularly acute in arid regions or where freshwater is supplied primarily from groundwater resources. Sparing water for other uses by adopting conservative water use in aquaculture is therefore a social obliga-tion. Water use is also inextricably tied to issues of waste discharge because water input and discharge are variables on two sides of the hydrological equation: increasing water input volume will increase water volume discharged.

Water use in aquaculture may be classifi ed as either total use or consumptive use (Boyd 2005). Total water use is the sum of all inflows (precipitation, runoff, seepage inflow, and management additions) to production facilities. Depending on the type of culture system, a portion of the water entering the facility passes downstream in overflow or intentional discharge and is available for other purposes (e.g., ecosystem support, irrigation). For example, essentially all water flowing into flow-through systems is discharged, whereas some ponds have no discharge at all. Consumptive use can be defi ned as water used in aquaculture that is not available for other purposes. Consumptive water use includes water lost by evaporation and seepage from an aquaculture facility, withdrawal of fresh ground-water, and water removed in biomass of aquatic animals at harvest (Boyd 2005). Water in harvest biomass averages about 0.75 L/kg, a minor quantity compared to other uses.

The requirement for water by aquaculture varies depending upon the type of culture system. Flow-through systems (such as raceways) require large volumes of water per unit fish produced. For example, water volume use in raceway culture of rainbow trout in the United States is approximately 98 m^3/kg (Table 1.3). The most common water supplies for flow-through systems are artesian springs or surface waters diverted from streams or rivers. Pumping water from wells or other sources is too expensive for large-scale com-mercial aquaculture and is seldom used except in small public hatcheries or in commercial hatcheries that are used for only part of the year (as in channel catfish farming, for example; Tucker 2005). Water flowing into culture units provides dissolved oxygen and water flowing out carries away metabolic wastes, such as carbon dioxide, ammonia, and fecal solids. Flow-through systems are usually confi gured so a given parcel of water is used more than once as it flows through a series of culture units. However, without treatment of the water to add oxygen or remove products of fish metabolism, there is a limit to the number of raceway segments that can reuse the same parcel of water.

Although flow-through systems require large volumes of water, they do not consume water unless it is pumped from aquifers. Except for negligible amounts of water lost to evaporation and fish harvest, inflow equals outflow. Boyd (2005) estimated that consump-tive water use in a typical trout flow-through facility is about 0.03 m^3/kg of production, or less than 0.05% of total water use. The common interpretation of water conservation is, therefore, meaningless in flow-through aquaculture because essentially no water is lost. However, efficiency of water use can be improved by using technologies that increase fish production per unit water flow. Although modern waste-management technologies used in flow-through aquaculture can remove a large proportion of the wastes added to water during culture, it is inevitable that some waste is discharged in flow-through system effl u-ents. In that regard, flow-through systems "use" water because downstream ecosystems are impacted and "appropriated" for waste treatment.

Water use in net-pen culture is similar to that in raceways. Adequate water quality within net pens depends on the exchange of water from outside the unit by tidal currents or river flow. In essence, net pens are simply flow-through systems constructed in water instead of on land. Like flow-through systems, large quantities of water must flow through the culture units to provide adequate dissolved oxygen and remove wastes, but net-pen facilities do not consume water. Most of the waste produced during culture is discharged directly into public, multiple-use water bodies and, as with raceways, net pens appropriate ecosystem area for waste treatment. The impact of appropriating surrounding areas for waste treatment can be minimized through proper site selection (see Chapter 8).

In contrast to flow-through and net-pen systems, many recirculating aquaculture systems are operated with very low water exchange rates. Water within culture units is reused many times by cycling water through mechanical and biological treatment processes to add dissolved oxygen and remove solids, ammonia, and dissolved carbon dioxide. In the extreme, water is used only to replace evaporative losses (which are small) and to replace water lost when concentrated waste solids are discharged from the system. Total and consumptive water use in recirculating can be less than 0.1 m^3/kg of production (Timmons et al. 2002).

Water use in ponds is functionally similar to that in recirculating systems in that water quality is maintained by mechanical and biological processes. However, in recirculating systems, treatment takes place in discrete units, such as screens, fi lters, or settling basins, whereas in ponds the processes are inherent parts of the ecosystem. Water use in ponds varies over a much wider range that for any other aquaculture system, depending primar-ily on the frequency of intentional water exchange (fl ushing) and pond drawdowns for harvest. At one extreme, water exchange in some ponds is so frequent that the systems operate hydrologically more like flow-through systems than ponds. In the 1980s and 1990s, marine shrimp ponds were routinely operated with daily water exchange rates of 10 to 20% of pond volume, and total water use (40 to 80 m^3/kg of shrimp produced) (Yoo and Boyd 1994) approached that of flow-through systems. Water used for water exchange is pumped from bays or estuaries and discharged back into the same water body, so only water lost to evaporation and seepage is used consumptively. More important than water use alone, however, water exchange generates large volumes of waste, increases the risk of pathogen transfer to and from the outside environment, and increases production costs associated with pumping. Eliminating or reducing water exchange by using water-reuse technologies can reduce total water use in shrimp ponds to less than 2 m^3/kg (see Box 7.3 in Chapter 7). Total water use in channel catfish farming (as an example of pond aquaculture in general) varies greatly depending on water source and phase of culture (see Chapter 6). Ponds may be filled with either watershed runoff or from wells, and consumptive use is much greater for ponds using groundwater. Consumptive use also varies among the dif-ferent phases of catfish culture because fi ngerling ponds are drained and refilled each year whereas brood ponds and foodfish ponds are operated for several years without draining. As such, more water is used for the fi ngerling phase because water drained each year must be replaced. Average consumptive water use in the foodfish-production phase is relatively low (less than 2 m^3/kg) (Hargreaves et al. 2002) because ponds are operated for long periods without draining. However, actual water use to produce foodfish must include water used in fi ngerling and brood phases of culture. When annual water use in the three culture phases are added and then divided by total annual foodfish production, average consumptive water use for pond-grown catfish in the southeastern United States is approx-imately 3 m^3/kg.

Consumptive water use is much higher in ponds than for other culture systems, and significant opportunity exits for water conservation. Considerable attention is therefore given to hydrology and water conservation in Chapters 6 and 7. Additional information on pond hydrology and water budgets is available in Yoo and Boyd (1994), Boyd and Tucker (1998), Hargreaves et al. (2002), and Boyd (2005).

The relatively high economic value of water used in aquaculture (also called the *con-sumptive water value index*) (Boyd 2005) can be shown by comparing consumptive water use and gross economic value for aquaculture and irrigated terrestrial crops grown in the same region. Using data from the southeastern United States, channel catfish aquaculture requires more water than irrigated cotton, corn, and soybeans, but an amount comparable to rice (Table 1.4). However, the value of various crops per unit volume of water used is much greater for catfish than other crops. If this is interpreted as an index of water use economic efficiency, catfish aquaculture is a more effi cient user of water than irrigated row crops.

Feed

The goal of aquaculture is to manage a body of water so that it will produce more fish, crustaceans, mollusks, or other animals than it would without management. Except for molluscan shellfish culture, this is accomplished by concentrating resources to provide food for the animal. There are two ways to provide food for aquatic animals: 1) fertilize an outdoor system to enhance natural food production and 2) grow foods outside the culture unit. The energetics and land-use issues associated with these two approaches were discussed above.

In the context of the ecosystem-level support for food production, the common aqua-culture systems used in the United States range from open-water mollusk culture, with no intentional resource input to enhance food supply, to net-pen and flow-through system culture of salmonids, where animal growth is totally dependent on resource expenditures outside the culture unit. Between those extremes is a continuum of food-supply strategies. In addition to molluscan shellfish culture, several other species are grown with relatively low resource input (aside from labor). For example, crawfish ponds operated at low inten-sity depend on development of a detrital food web based on natural plant forage. Some baitfish and ornamental fish ponds are also operated at relatively low intensity and are fertilized to promote an abundant supply of natural foods. However, more than 95% of all commercial finfish aquaculture and 70% of total aquaculture production in the United States depends on manufactured feeds. Worldwide, manufactured feeds support approxi-mately 40% of all animal aquaculture production (FAO 2006b).

Manufactured feeds (Fig 1.7) are formulated to provide all essential nutrients and energy needed to promote rapid growth of healthy animals. Diet composition varies widely (Table 1.5) depending on the nutritional requirements of the species cultured, cost of feedstuffs, and the relationship of feed formulation to body composition, disease resistance, and waste production. Feeds are typically formulated from a mixture of animal and plant feedstuffs (Lovell 1998). The major animal feedstuff in most aquafeeds is fishmeal derived from pelagic marine feed fish such as anchovy, menhaden, and herring. Other animal feedstuffs include poultry and beef processing by-products. The primary plant feedstuffs are meals or by-products derived from soybeans, wheat, corn, canola and rapeseed, rice, sorghum, cotton seed, or peanuts. Diets may also contain various animal and plant fats or oils to increase feed energy content, provide essential fatty acids, and reduce dust or fi nes pro-duced during feed handling. Fish oil is presently an indispensable ingredient in diets for salmonids and other carnivorous finfish and is a coproduct of fishmeal production. Feeds may also contain supplemental vitamins and minerals, binding agents, or pigments. Summaries of feed formulation for most aquaculture species are provided in Lovell (1998) and Webster and Lim (2002).

Environmental issues associated with manufactured feed

Two environmental issues have become associated with using manufactured feeds in aquaculture. First, wastes derived from uneaten feed and excreted nutrients are a potential source of pollution if culture water is discharged from the facility. Relationships among waste production, feed composition, and feeding practices are therefore critical to manag-ing potential impacts of aquacultural effl uents (Gatlin and Hardy 2002). Better manage-ment practices for feeds and feeding are included in each of the "systems" chapters of this book (Chapters 6 through 10).

In addition to the relationship between feed and waste production, additional concerns have been raised about the environmental impacts of procuring ingredients for aquaculture feeds. In the broadest sense, the concern is that procuring and manufacturing aquaculture feeds have impacts far beyond the culture facility. Two of these issues have already been discussed. First, feedstuff procurement, manufacture, and other associated activities consume the largest proportion of energy used in producing fish or crustaceans that rely on manu-factured feeds (Fig 1.8). Second, land and sea area is appropriated from outside the facility to supply feedstuffs that are then processed and combined in a manufactured feed that is used to grow confined aquatic animals. Depending on the species cultured and the nature of the production system, the external ecosystem support area for food production may range from near zero to many thousands of times larger than that of the facility itself. Questions associated with large facility "footprints" of intensive aquaculture of carnivorous species include whether a given ecosystem can indefinitely provide resources to support aquaculture production and whether the appropriated resources have greater social or eco-nomic value with some other use. These questions are most often debated in the context of fishmeal. A larger ecosystem support area is required to produce fishmeal than terrestrial feedstuffs, and fishmeal may become limiting to expansion of certain forms of aquaculture because there is little prospect for increased global supply from marine capture fisheries.

The fishmeal issue

Fishmeal is produced by cooking, pressing, drying, and milling whole fish, fish scraps, or fish-processing wastes. Fish oil is a valuable coproduct of fishmeal production. Fish and crustaceans do not have a dietary requirement for fishmeal or fish oil. Rather, aquatic animals require specific nutrients contained in the meal or oil, and in the past they were an economically favored feedstuff to provide those nutrients. Fishmeal is an excellent protein feedstuff, especially for aquaculture, because it has a well-balanced amino acid profi le that closely matches the dietary requirements of carnivorous marine finfish, is highly palatable, is highly digestible, and until recently, was a relatively inexpensive protein source. Fish oil has historically been an economical source of long-chain, highly unsaturated fatty acids that are essential in the diet of carnivorous fish and crustaceans. Fish oil, as part of aquaculture feeds, is also useful in manipulating the tissue fatty-acid profi le to provide higher levels of the omega-3 fatty acids eicosapentaenoic acid (EPA) and docosahexaenoic acid (DHA). Adequate dietary intake of omega-3 fatty acids by humans has been shown to reduce the risk of cardiovascular diseases and is a highly pub-licized benefit of increasing the consumption of seafood rich in these fatty acids.

About 85% of the global fishmeal supply is derived from capture of pelagic fish, primar-ily from nutrient-rich, oceanic upwelling areas of continental margins. The remainder is derived from fish scraps and processing waste (Tacon et al. 2006). Roughly 30 million tonnes of fish, representing about 23% of global fisheries landings, were reduced to about 6.5 million tonnes of fishmeal and 1 million tonnes of fish oil in 2004 (FAO 2006b). The capture of fish for reduction to fishmeal has been stable since the mid-1980s, except when Peruvian and Chilean anchovy fisheries decline precipitously from disruptions to upwell-ing water currents known as the El Niño Southern Oscillation. Most reduction fisheries stocks are considered fully exploited, with some stocks overexploited and some moder-ately exploited. There is no reason to expect increases in the global fishmeal supply.

Although global fishmeal supplies have been static since the 1980s, increased demand has caused fishmeal prices to rise. Among various users of fishmeal, the aquaculture sector has been more willing to pay the increased price and the proportion of fishmeal used by different agriculture sectors has changed dramatically. In the early 1990s, aquaculture used 10 to 15% of the world's fishmeal supply; the remainder was used primarily in swine and poultry production. In 2005, aquaculture used 53% of global fishmeal supplies and 87% of fish oil, despite aquaculture representing only 3% of total industrial animal feed produc-tion (Tacon et al. 2006).

Aquaculture is projected to consume an even larger proportion of the global fishmeal supply as farming of shrimp and carnivorous fish expands. Marine shrimp and marine finfish represent about 25% of global aquaculture production, but consume 75% of the fishmeal and fish oil used in aquaculture feeds (Tacon 2004). Also important is the large-scale change in Asian aquaculture, particularly in China, where traditional extensive pond culture of carp is rapidly giving way to carp aquaculture based on manufactured feeds containing fishmeal (Tacon et al. 2006). The potential demand for fishmeal by this aquaculture sector is immense.

Using fish in feeds for other fish has raised a concern that is often cited as possibly the most significant environmental impact of aquaculture: that is, the culture of fish and crus-taceans on high-fishmeal feeds does not contribute to an overall increase in global fish production. In other words, each kilogram of farmed fish consumes more than 1 kg of feed fish in the form of fishmeal (Naylor et al. 1998, 2000, 2001; Naylor and Burke 2005).

The amount of feed fish needed to grow a kilogram of fish or crustacean in aquaculture (called the *feed-fish equivalence*, or *FFE*) is calculated from the feed conversion ratio achieved during culture (FCR = weight feed fed/weight fish produced), the percentage fishmeal in the feed, and the ratio of the quantity of feed fish required to produce a unit of fishmeal. The weight of feed fish required to produce a unit weight of fishmeal varies among species, but usually ranges from 4 to 5 (Tacon et al. 2006). Assuming that 4.5 kg of feed fish is reduced to 1 kg of fishmeal, the feed-fish equivalence is calculated as follows:

FFE = [(% fishmeal in feed x 4.5) x FCR] -s- 100 (1.1)

For example, assume 1 kg of Atlantic salmon is grown with a feed containing 35% fish-meal at a feed conversion of 1.2. The feed-fish equivalence is 1.9 kg of feed fish from pelagic fisheries. Channel catfish cultured with a feed containing 2% fishmeal and a feed conversion of 2.2 would require 0.2 kg of feed fish for every 1 kg of catfish produced. Based on these two examples, it could be argued that Atlantic salmon aquaculture reduces overall global seafood production and that channel catfish production is a net contributor. Species groups that are net consumers of fish include river eels, salmon, marine finfish, trout, and marine shrimp. Species groups that are net producers of fish include carp, catfish, tilapia, milkfish, and freshwater crustaceans (Tacon et al. 2006). Globally, 20 to 25 million tonnes of feed fish were used in aquaculture to produce 30 million tonnes of fish and crustaceans in 2003 (Tacon et al. 2006).

The concern with "fish eating other fish" is that, for some species, it appears to be ecologically ineffi cient (Box 1.2). Superfi cially this is true, but it neglects other consider-ations (Tidwell and Allan 2002). Feed fish and high-valued fish and shrimp produced in aquaculture do not substitute equally either socially or economically, so accounting based on "fish-in versus fish-out" is too simple to assess the deeper signifi cance of fishmeal use in aquaculture. Most fish caught for reduction to fishmeal are not desirable as human food in many countries. Using fishmeal in aquaculture feeds adds value by converting low-value protein not used for human food into a high-value product. The additional steps in the process of using feed fish for animal food rather than human food also creates additional economic opportunities, such as jobs in the feed manufacturing and aquaculture sectors. The apparent ecological inefficiency of the process is also misleading because ineffi cien-cies are inevitable in nature when organisms at one trophic level consume biomass from a lower trophic level. In fact, some have argued (Asgard et al. 1997; Forster 1999) that it is considerably more effi cient to capture, process, and feed pelagic fish to salmon than to allow the fish to be eaten naturally by wild fish where ecological ineffi ciencies are on the order of tenfold for each change in trophic level. Perhaps the most important consideration is that aquaculture has not caused greater overall use of fishmeal. Global fishmeal produc-tion and use has been relatively stable since the 1980s and has been redistributed only among various users, with aquaculture using a larger proportion over time.

Although some forms of aquaculture are, as described above, net consumers of fish, it is worth remembering that fishmeal use in poultry and swine production also does not add to global seafood supplies. The argument might be made that fishmeal use in feeds for certain terrestrial animals is a better use than in aquaculture because dietary inclusion rates are low and the net result is an increase in global supplies of edible animal protein. The same argument could, however, be made for using fishmeal in feeds for fish—such as channel catfish, tilapia, and carp—that also have low fishmeal inclusion levels in feeds.

Trends in mean aquaculture trophic level vary greatly among countries and regions. The mean trophic level for species grown in Canada, Chile, Norway, and the United Kingdom increased rapidly—more than one whole trophic level unit—from 1980 to 1997 (Pauly et al. 2001a, b). This corresponds, of course, to expansion of salmon aquaculture in those countries. Over the same period, the mean trophic level for species grown in China remained essentially unchanged and those grown in the United States decreased slightly. In fact, of the eight countries analyzed, the United States had the lowest aquaculture mean trophic level. This was attributed to the dominance of United States aquaculture by channel catfish, a species grown on feeds with low levels of fishmeal inclusion. Since 1997 (the last year in the analysis by Pauly et al. 2001b), increased usage of fishmeal in feeds for carp and other species in China (Tacon et al. 2006) has probably caused increases in the trophic level index for that country. Meanwhile, catfish aquaculture continued to expand and fishmeal levels in catfish feeds have decreased even further since 1997. Also, other sectors of low trophic level aquaculture—such as freshwater crawfish and molluscan shellfish— have contributed significantly to United States aquaculture, and the mean trophic level for cultured species has undoubtedly continued to decrease.

Many of the potential environmental impacts of aquaculture are localized in the vicin-ity of aquaculture facilities. In contrast, the use of fishmeal and fish oil in aquaculture is an issue that is global in scope, affecting all sectors that are supported by manufactured feeds. The main environmental and social concerns associated with fishmeal and fish oil use in aquaculture include

- Fishmeal and fish oil resources are limited. As discussed above, the yield from reduc-tion fisheries for small pelagic species has been static since the 1980s and further increases in supply are highly unlikely. This has obvious implications for the sustain-ability of different forms of aquaculture, particularly those dependent on fishmeal as a protein source in manufactured feeds. Allocation of this finite resource among various forms of animal protein production is likely to become much more competitive in the future.
- Reduction fisheries are currently managed to maximize long-term yield, without con-sideration of wider ecosystem effects. Important among these impacts are the trophic interactions that occur in ecosystems in which small pelagic species support a fishery for carnivorous species or function as food resources for marine species with inherent value with respect to biodiversity, but without commercial value.
- Some of the species that are currently harvested and reduced to fishmeal are suitable for direct use as human food. Reducing these fisheries resources to fishmeal, incorporating into aquaculture feeds, and feeding to marine shrimp and finfish raises ethical issues about allocation of global supplies of animal protein and effects on human well-being.

Fishmeal and fish oil use in aquaculture may be a self-limiting issue due to market factors. Fishmeal prices, which have doubled since 2005, will continue to rise as increasing demands are made on the finite fishmeal supply. Contributing to the price increase will be steadily increasing fishing fl eet fuel costs. Under the scenario of higher fishmeal prices, there will be increased incentives to grow aquatic animals with lower dependence on fish products in manufactured feeds and to develop dietary substitutes for fishmeal and fish oil in aquaculture feeds, which remains an active area of research. There is consider-able scope for the reduction of fishmeal and fish oil content of aquaculture feeds for marine finfish. There has been some progress in replacing fishmeal with oil-seed meals, although much less progress in replacing fish oil. Forms of aquaculture that continue to depend on high levels of fishmeal inclusion in feeds may become uneconomical.

There is a significant environmental trade-off regarding fishmeal in aquaculture feeds. Regardless of the ethics or economic implications of using fishmeal and fish oil in aqua-culture feeds, incorporating them into feeds at appropriate inclusion levels can be one of the most effective strategies for reducing waste production in aquaculture. The use of highly digestible, high-energy feeds is especially important when fish are grown in open aquaculture systems such as raceways and net pens (see Box 9.1 in Chapter 9). Improvements in feed formulation can improve the efficiency of feed utilization and thereby reduce the discharge of waste nutrients. However, reducing the demand for fish-meal and fish oil from aquaculture would not necessarily reduce pressure on reduction fisheries because these products are globally traded commodities.

POTENTIAL ENVIRONMENTAL IMPACTS OF AQUACULTURE

The environmental effects of aquaculture are often described in terms of the sources (type and quantity) or causes of a potential effect, and this convention will be followed here. Impacts are potential because effects depend on characteristics of the environment in which aquaculture production systems are embedded. For example, releasing waste into a closed basin will have a qualitatively greater (more adverse) impact that releasing the same quantity and type of waste into an open basin.

Contrasting to broad issues related to resource use, many of the problems commonly attributed to aquaculture—such as the adverse effects of habitat conversion, waste dis-charge, consumptive water use, and release of chemicals and antibiotics—are related to facility operations and can be addressed on shorter time scales through technology-based approaches (*better management practices,* or *BMPs*). Solutions to problems of this nature are the focus of this book.

Habitat Conversion

Ponds are the most common aquaculture production system in the United States and the world. In most cases, pond construction requires conversion of terrestrial or wetland habitat to aquatic habitat. Aquaculture ponds must be located near sources of water, so they are commonly constructed in coastal or riverine fl oodplains. Thus, pond construction may convert fl oodplain land or wetlands to ponds (Fig. 1.9).

Pond construction in or near coastal wetlands, particularly the conversion of tropical mangroves to shrimp ponds, is controversial. Coastal wetlands are multiple-use, mostly open-access resources that are increasingly threatened by a wide range of development pressures, including aquaculture. Coastal wetlands provide a number of ecosystem services, including waste treatment, storm protection, food production (especially as a nursery for many aquatic species), recreation, and a source of raw materials. The environ-mental impact of pond construction in coastal wetlands can be described in terms of the degree to which these services are lost by conversion to ponds. The total economic value (including the value of nonmarketed ecosystem services) of sustainably managed man-grove wetlands often exceeds the value associated with conversion to shrimp ponds (Rönnbäck 1999; MEA 2005). Coastal development, including aquaculture, in tropical areas can adversely affect adjacent, light-sensitive ecosystems, such as coral reefs and seagrass meadows.

Approximately 35% of mangroves have been lost in the last 2 decades (MEA 2005). Estimates of the relative contribution of aquaculture to the loss of coastal mangroves are uncertain, although less than 10% of the global loss of mangrove area has been attributed to conversion into shrimp ponds (Macintosh and Phillips 1992; Boyd and Clay 1998; World Bank et al. 2002). Nonetheless, conversion to aquaculture ponds can be the major component of mangrove loss in some coastal areas, particularly in southeast Asia (Primavera 2006).

Mangrove wetlands are increasingly seen as having intrinsic value, being poor sites for pond construction, and undesirable sources of social conflict when aquaculture projects are developed. Pond aquaculture in coastal wetlands is much less common than in the past. In many places in Asia, mangrove clearing has been banned and rehabilitation and restoration programs are active.

Waste Loading

Nearly all aquaculture production systems produce waste nutrients and organic matter. Depending on the production method, organic fertilizers (manures and agricultural by-products), inorganic fertilizers, and nutrient-dense feeds are applied to production units in most aquaculture facilities to promote growth of finfish, crustaceans, and other species. Approximately 70 to 80% of nutrients in feed are released as waste into the culture unit in aquaculture production systems where cultured animals are fed.

The degree to which the culture unit is open to the environment will determine the proportion of waste excreted by cultured animals that is released to the environment. For example, ponds and recirculating systems are operated with a long hydraulic residence time, so a very small proportion of nutrients are released to the environment. In contrast, culture systems with a more direct hydrological connection with the outside environment, such as flow-through systems and net pens, will release a large proportion of waste nutri-ents and organic matter.

The ecological effects of discharged nutrients will vary depending on production inten-sity, facility density, the hydraulic retention time and trophic state of receiving waters, and dilution rate. In relatively closed production systems such as ponds, feeding invariably results in the eutrophication of pond water. In more open systems, the effects of nutrient discharge are far less predictable. If the fl ushing time of the water body is less than the generation time of phytoplankton, an increase in limiting nutrient concentration can result in eutrophication. It is important to emphasize the point that phytoplankton respond to changes in nutrient concentration, particularly of those nutrients that tend to limit produc-tion (phosphorus in freshwater and nitrogen in seawater), not to increases in nutrient mass loading per se.

Potential adverse environmental effects attributed to organic waste loading from aqua-culture include increases in phytoplankton density in response to elevated nutrient con-centration, increased variation in dissolved oxygen concentration in response to increased phytoplankton density, increased frequency of dissolved oxygen depletion, reduced dis-solved oxygen concentration caused by the discharge of effl uent with high biochemical oxygen demand, and changes to benthic communities caused by localized sedimentation of suspended solids. Nutrients discharged from aquaculture facilities have not been linked to the development of harmful algal blooms.

In many coastal areas, discharges from aquaculture production facilities represent one of many nutrient sources contributing to eutrophication. Truly addressing problems related to eutrophication therefore requires that the relative contribution of aquaculture be evalu-ated in terms of all sources and considered in the context of a watershed or whole-basin approach (Boyd and Green 2002). In certain locations, particularly closed basins, aqua-culture operations can be the primary contributor to localized eutrophication. The effects of effl uent discharge tend to be localized near the discharge point. The susceptibility of receiving waters to eutrophication depends on trophic status prior to enrichment. For a given level of nutrient loading, oligotrophic waters are more susceptible to trophic state changes than mesotrophic waters.

Most discussions of waste loading from aquaculture consider only the components of the waste stream, which includes nutrients, organic matter, potential pathogens, chemicals or therapeutants, and escaped fish. However, they do not consider how the environment responds to waste loading. Often missing is the notion of environmental carrying capacity, which requires an understanding of the waste assimilation capacity of the environment and consensus on an acceptable level of environmental change.

Aquatic environments have an inherent capacity to assimilate nutrients through well-known biological and physicochemical processes. In some areas, aquaculture facilities are sufficiently concentrated and production intensity is sufficiently great that the majority of the assimilative capacity of receiving waters is used by aquaculture.

Appropriation of the waste treatment ecosystem service by aquaculture has been criticized on the basis that the cost of waste treatment is externalized by aquaculture operations and born by society. In general, nutrient enrichment is perceived to have negative consequences, although in some environments, eutrophication may be beneficial and perceived as desirable. Changes in trophic state have no inherent value; the acceptability of the change is a matter of societal values and policy.

Benthic Impacts

Organic wastes produced in open aquaculture production systems can accumulate on the sediment in the vicinity of culture units. Soil eroded from pond embankments and organic matter in pond water can settle in receiving water bodies adjacent to discharge points. Organic matter discharged from flow-through systems can accumulate near the discharge point, leading to localized suppression of dissolved oxygen concentration and changes in benthic communities. Depending on current speed and water depth, organic matter mainly derived from feces and uneaten feed can accumulate on the sea fl oor beneath net pens. Sedimentation rates within net-pen facilities are reported to range from about 15 to 100 g of total volatile solids (TVS)/m^2 per day, with most estimates between 25 and 50 g TVS/m^2 per day (Brooks and Mahnken 2003a). The accumulation of organic matter can extend from 145 to 205 m downcurrent from the perimeter of net-pen facilities (Brooks and Mahnken 2003a), although significant effects are usually restricted to less than 60 m from the perimeter (Nash et al. 2005). Of all the potential environmen-tal impacts of salmonid net-pen aquaculture, changes to the sediment beneath net pens are considered to represent the greatest risk to the environment (Brooks and Mahnken 2003a; Nash et al. 2005), although these changes are temporary and largely reversible.

If the rate of organic matter sedimentation exceeds the rate of decomposition, organic matter will accumulate. Organic matter accumulation will depend on loading rate, tem-perature, oxygen supply, current velocity, and other physical and chemical factors. The deposition of organic matter creates a dissolved oxygen demand for decomposition that may exceed supply. In this case, the redox potential will shift to lower (more negative) values, indicating anaerobic conditions. Mats of the sulfi de-oxidizing fi lamentous bacteria *Beggiatoa* may cover the sediment surface. Reduced substances such as ammonia, dis-solved phosphorus, methane, carbon dioxide, and especially in marine waters, hydrogen sulfi de will diffuse into the overlying water. Some of these substances are potentially toxic to benthic invertebrates and cultured fish, but rarely accumulate to levels that will affect fish culture performance.

Anaerobic conditions in the sediment will also cause a shift in the assemblage of benthic invertebrates toward a less diverse community dominated by pollution-tolerant species. The effects described here tend to be localized around effl uent discharge points and within 25 m of the perimeter of net-pen farms (Karakassis et al. 2000; Pearson and Black 2001). These effects are also more pronounced for poorly fl ushed sites (current velocity <10 cm/ second) with fi ne-grained sediments. In well-fl ushed sites (>50 cm/second), the abundance and diversity of benthic infauna frequently increases (Brooks and Mahnken 2003a). Chemical remediation of marine sediments beneath net pens occurs within months to 1 year, but biological remediation may require 2 to 3 years in some sites, largely depending on water temperature, current velocity, and concentration of sediment organic matter. In general, the effects of sedimentation are localized and reversible.

Sedimentation can also occur as a result of alteration in **hydrodynamics** caused by the deployment of aquaculture structures such as rafts, longlines, and cages. These can reduce current velocity in the lee of culture facilities, creating a depositional environment. In addition, feces and psuedofeces produced by shellfish can accumulate in the vicinity of shellfish rafts. Hydrodynamics can be altered by material that accumulates near effl uent outfalls, which can reduce stream depth and obstruct water flows, increasing the potential for fl ooding.

Some metals will accumulate in sediment, but there is no evidence that these accumulations have had negative impacts on benthic invertebrates or rates of biogeo-chemical transformations (Brooks and Mahnken 2003b). Zinc is added to fish feeds as an essential nutrient, and elevated concentrations have been measured in sediment beneath net pens. Copper is used in antifouling paints and net treatments and can accumulate in sediment beneath net pens. In marine environments, the accumulation and bioavailability of copper and zinc depend on sulphide concentration, with reduced bioavailability with elevated sulphide concentration. Concentrations decline during site fallowing.

Chemical Pollution

As with other forms of agriculture, chemicals are used in aquaculture for a broad diversity of purposes. Perhaps most importantly, pesticides, disinfectants, and antibiotics are used for disease treatment and pest management. Chemicals are also used for soil and water treatment, enhancing natural productivity, feed manufacture, control of reproduction, growth enhancement, transportation, and processing. Major classes of chemicals include therapeutants (including antibacterial agents), disinfectants, anesthetics, hormones, feed additives, antifouling paints, herbicides, and pesticides (GESAMP 1997; Boyd and Massaut 1999; Gräslund et al. 2003). Soil and water are treated with chemicals such as inorganic and organic fertilizers, liming agents, fl occulating agents, and bacterial amendments. Chemical and drug use are highly regulated in the United States (as summarized in Chapter 12), and even usage of legally registered substances is relatively uncommon for reasons of economics. This is not necessarily true elsewhere, however. On average, a shrimp farmer in Thailand uses 13 chemicals, with four pesticides and disinfectants, and three soil and water treatment products (Gräslund et al. 2003). Thai shrimp farmers use an average of one antibiotic per farm, most commonly fl uoroquinolones.

Chemicals used in aquaculture can be divided into three groups based on the different ways they affect the environment (Gräslund and Bengtsson 2001). One group of chemicals encompasses those with acute or chronic toxicity to nontarget organisms, including cul-tured animals. Another group are the antibiotics that are released to the environment. These can lead to the development of resistant strains of pathogenic bacteria and alter microbial community composition, affecting biogeochemical processes. The third group are nutri-ents that can lead to eutrophication if discharged; a particular concern for light-sensitive seagrass and coral reef ecosystems often found adjacent to coastal shrimp ponds. Chemicals can also be grouped into three categories based on the environmental or human health risk associated with their use (GESAMP 1997). The fi rst group includes chemicals with a high risk associated with their use, such as chloramphenicol, malachite green, and organophosphates. Chemicals in a second group can be used safely if applied according to label directions, but may present an environmental or human health hazard if used incorrectly. A third group of chemicals can be used safely at most locations, but may present an elevated risk at certain sites because of particular characteristics of that site.

Chemicals have various fates when released from aquaculture production systems to the environment, so environmental persistence is variable. The persistence of chemicals in the environment depends on physicochemical factors that affect solubility and reactiv-ity (especially temperature, pH, dissolved oxygen concentration, and light intensity) and biological factors, especially the type and density of microorganisms. Some chemicals can accumulate in sediments. Other chemicals dissipate or degrade by physicochemical reac-tions such as photooxidation or adsorption to sediment minerals or organic matter. Others are transformed or degraded biologically by reactions mediated by microorganisms. The environmental toxicity of reaction products can be greater or less than that of reaction substrates. In general, but with some important exceptions, most chemicals used in aqua-culture do not persist in the environment (Gräslund and Bengtsson 2001).

The environmental and human health issues associated with chemical use in aquaculture were summarized by GESAMP (1997). These include persistence in aquatic environments, residues in noncultured organisms, toxicity to nontarget organisms, stimulation of bacterial resistance, effects on sediment biogeochemistry, nutrient enrichment, health of workers exposed to chemicals, and residues in seafood that affect product safety. Many of these effects can disturb natural aquatic communities, altering community structure and reduc-ing biodiversity.

There is a strong disincentive to use many chemicals in aquaculture because use increases production costs and some chemicals may be toxic to cultured aquatic animals. Chemical use also presents a food safety problem by increasing the risk of consumption of aquatic animals containing chemical residues. Improper use of antimicrobials can stimulate the development of drug-resistant forms of microbial pathogens in cultured animals, water, or sediment. Chemicals can also bioaccumulate in nontarget organisms and become biomagnifi ed as they move through aquatic food webs to higher trophic levels.

In the particular case of salmonid net-pen aquaculture, there has been concern about the use and fate of pesticides and drugs applied to control infestations of various species of parasitic copepods known collectively as sea lice and the antibiotics used to control bacterial diseases. In some countries, sea lice are controlled with bath treatments of nonspecific and broad-spectrum organophosphorus and pyrethroid pesticides and with the antibiotic ivermectin incorporated into feed. The chemicals used in bath treatments disperse rapidly and are usually not detected beyond 25 m from a net pen, although these chemicals have the potential to affect nontarget organisms, particularly crustaceans. Research on the fate of antibiotics used to control bacterial diseases indicates that antibi-otic residues will accumulate and persist in sediment beneath cage sites and have been measured in nontarget organisms (Samuelson et al. 1992; Smith et al. 1994; Weston 1996; GESAMP 1997). There is concern about the increased risk of the development of drug-resistant strains of bacteria that are pathogenic to fish (Samuelson et al. 1992), particularly when antibiotics are used prophylactically. Similar concerns have been expressed about the development of drug-resistant strains of bacteria in shrimp farming (Gräslund and Bengtsson 2001; Holmström et al. 2003).

Salinization

Salinization refers to increases in the salinity of soil, surface water, and groundwater. Excessive salinization of water and soil can limit the use of surface waters for irrigation, reduce crop yields, displace salt-intolerant crops, and limit the use of water for domestic purposes. Salinization is primarily a concern with coastal pond aquaculture, where water is pumped to maintain appropriate salinity in ponds for penaeid shrimp culture. However, salinization has become a concern with the increase in inland aquaculture of marine shrimp. These facilities are supplied with brine transported from coastal waters or saline groundwater from wells, and salinization can occur when this water is discharged to inland streams. Salinization can also occur when accumulated sediment is removed from coastal aquaculture ponds and disposed in freshwater areas. Salt contained in the sediment leaches out following rainfall and can cause salinization (Boyd et al. 1994).

The discharge of water from coastal or inland brackishwater ponds into freshwater bodies can impair surface water quality to limit their use, especially for agriculture (Dierberg and Kiattisimkul 1996; Braaten and Flaherty 2001; Boyd et al. 2006). In sites with improperly compacted soils, water can seep from these ponds into underlying fresh-water aquifers. Alternatively, excessive pumping of groundwater can result in saltwater intrusion of aquifers, thereby limiting use. Excessive pumping of groundwater to control salinity in coastal aquaculture ponds has led to salinization of groundwater aquifers and land subsidence and consequent damage to infrastructure in Taiwan (Lin 1989; Huang 1990). Pumping groundwater was widely used to manage salinity in brackishwater ponds in the past, but the practice is now banned or restricted in many places.

Pathogen Transmission

In aquaculture production systems, animals are usually held at much greater densities than in the natural environment, a condition that facilitates the transmission of pathogens among cultured animals. Elevated production intensity can lead to environmental conditions in the production unit that stress cultured animals, potentially increasing the susceptibility to infection by pathogens that may lead to disease. Thus, the conditions that promote disease are more prevalent in aquaculture production systems than in the natural environment. Intensive culture conditions can also cause amplifi cation of pathogens that are ubiquitous in the environment.

Pathogens are a constituent of water that is discharged from or flows through aquacul-ture facilities. Water exchange in open production systems facilitates the transmission of pathogens to the environment, specifically between cultured and wild fish. There are numerous examples of diseases caused by the discharge of pathogens from aquaculture production facilities. Wild fish also serve as a reservoir of pathogens for the infection of cultured fish. The transfer of pathogens to wild fish can potentially affect the susceptibility of wild fish to disease and, in extreme cases, could affect wild fish abundance and local biodiversity.

The discharge of pathogens does not necessarily lead to disease. Infection with a patho-gen is a much more common occurrence than disease. Infectious diseases result from the interaction of the health and immunological status of the host, the dose and virulence of the pathogen, and environmental conditions that affect host-pathogen interaction. The health of wild fish populations is influenced by natural factors, such as genetics, nutrition, environmental conditions, and the biology of pathogen and host organisms. Fish health is also affected by anthropogenic factors that put pressure on wild fish populations, including climate change, fl oods, drought, impoundments, dams, chemical contaminants, and fishing pressure.

The risk of disease caused by a particular pathogen is affected by the characteristics, distribution, survival, and fate of pathogens in the environment; factors affecting the route of pathogen transmission; the characteristics of pathogen exposure; the effect of environ-mental stressors; and the susceptibility of wild hosts to infection that can lead to disease (Blazer and LaPatra 2002). In general, the relative importance of each factor is difficult to understand and assess in wild populations, leading to a high degree of uncertainty associated with the risk of transfer of pathogens between cultured and wild fish popula-tions (Coutant 1998; Raynard et al. 2007).

Wild fish can be affected by the introduction of new pathogens attendant with the intro-duction of native or nonnative species to an area for culture. These pathogens can infect wild individuals of the same or different species. Pathogens can also be introduced to wild fish through transportation of native or nonnative species that are not intended for aquaculture.

Interactions of Escaped Fish with Wild Populations and Natural Ecosystems

Aquaculture production facilities are embedded in natural ecosystems with varying degrees of intimacy, depending on characteristics of the culture system. The hydrological connec-tion to natural ecosystems can be remote and intermittent (e.g., recirculating aquaculture systems) or intimate and continuous (e.g., shellfish and net-pen systems). The risks of containment failure also vary with culture system type. As such, the probability of occur-rence of escape is variable. Escapes can be defi ned as the unintentional and unplanned release of cultured animals or their gametes or offspring to the environment. Escapes do not include intentional releases of cultured animals for stock enhancement or ranching.

Mechanisms that lead to the escape of cultured organisms can be grouped into four classes (Myrick 2002). In each case, the mechanism leads to a breach in the barrier between a culture unit and the environment, allowing escapes to occur. The causes of containment failures that can lead to escape include 1) catastrophic natural events, 2) human error associated with facility operation, 3) vandalism or poaching, and 4) attacks by nuisance or predatory wildlife. Culturists have no control over natural disasters, but facilities can be constructed to withstand or minimize the risk of escapes by other mechanisms.

Escape of cultured animals is not desirable from an economic standpoint because escapes represent a loss of potential revenue. Therefore, there is a strong incentive to construct and manage production units to minimize losses from escapes, disease, poaching, and other causes. Above and beyond the undesirable economic impact on producers, escaped culture organisms can alter the physical environment and the structure of eco-logical communities, cause genetic changes in stocks of wild conspecifics, introduce parasites and diseases, and cause various socioeconomic impacts (Beveridge and Phillips 1993).

Escaped individuals may interbreed with natural populations, although interbreeding requires the satisfaction of several criteria. First, cultured organisms must be sexually mature or, if not mature, they must survive to maturity. The timing of the escape event must also coincide with the natural breeding season of the wild population; in other words, sexually mature individuals in the wild population must be receptive to escaped fish.

Intraspecific hybridization between escaped and wild conspecifics can reduce the genetic variance, and presumably fi tness, of wild populations through outbreeding depression (Kapuscinski and Brister 2001). This may increase the vulnerability of natural populations to environmental change from the loss of genetic differences between cultured and wild (natural) populations. The effect of inbreeding will depend on the degree of local adapta-tion. Geographically structured populations are more susceptible to the effects of interbreeding than undifferentiated stocks.

The principles of interaction between wild and escaped fish have been reviewed by Youngson et al. (2001). The overall concern is related to a reduction in fi tness and a reduc-tion in the effective population size of wild populations.

First, the genetic profi le (i.e., allele frequencies) of wild populations can change from the transfer of genes from cultured stocks into wild populations. Second, the genetic profi le of cultured stocks may contain alleles that are rare or unusual in wild populations but were selected on the basis of performance traits, and that are transferred to wild stocks. Third, the abundance of wild populations may decrease if escaped fish displace wild fish, but then demonstrate low reproductive fi tness. Although all of these mechanisms of interaction are plausible, dem-onstrating a measurable effect on the fi tness of wild populations is difficult.

The consensus view is that the risk of adverse genetic effects on natural populations of conspecifics is not as severe as potential ecological effects of escapes. The intensity of the ecological impacts of escapes is a function of the number of organisms that escape, the timing of an escape event, the biology and life history of the escaping species, interactions of escaped organisms with wild populations, and ecological characteristics of the host ecosystem. Even nonreproductive animals can have ecological impacts if they escape in sufficient numbers.

Potential adverse impacts of escapes on natural populations may include any or all of the following (Myrick 2002):

1) Competition with natural populations for habitat, food resources, nesting or spawning sites, or mates;
2) Predation on endemic fish populations if escaped fish are piscivores;
3) Predation on endemic fish populations if escaped fish attract predators;
4) Transmission of pathogens or parasites with deleterious effects on wild populations;
5) Amplifi cation of the effect of endemic pathogens;

6) Alteration of habitat, such as changes in turbidity and substrate; and
7) Colonization of habitat, with negative effects on wild populations, and in the worst case, displacement of native species.

Overall, the potential outcomes of escapes may include a reduction in fi tness in wild con-specific populations and a reduction in fi tness in other wild populations.

Many of the adverse impacts described above are exacerbated when invasive species escape from aquaculture facilities. Although only one of many pathways for the introduction of potentially invasive species, aquaculture has played an important role in the introduction of nonnative species, especially in freshwater. The majority of introduc-tions of nonnative species since the 1970s can be attributed to aquaculture (Welcomme 1988) and arguably the introduction and establishment of invasive species from aquacul-ture has had the greatest adverse effect on the environment of all potential impacts. In any case, the effects of escapes by cultured species on biodiversity are difficult to parti-tion from the effects of habitat degradation or loss and other causes of biodiversity reduction.

Collection of Wild Aquatic Animals for Spawning or Stocking

There is a long history in aquaculture of capturing wild larvae and juveniles for stocking into culture units for growth to market size. The culture of milkfish, mullet, penaeid shrimp, and many other species was long dependent on wild stocking material. Capturing wild juveniles is usually an interim step in the development of a species for culture. Once the techniques for spawning and early larval rearing are developed to close the life cycle, juveniles can be reliably produced in hatcheries, allowing the development of a sustainable aquaculture industry for that species. Most species in freshwater aquaculture are now spawned in captivity. The dependence on wild-caught stocking materials is becoming much less common, although reliance is still strong in brackishwater and marine aquacul-ture. The capture-based aquaculture for eels, groupers, tunas, and yellowtails remains dependent on capture of wild juveniles (Ottolenghi et al. 2004).

There are several problems related to the use of wild juveniles for stocking. Collection of juveniles may place additional pressures on wild stocks of the target species, reducing recruitment to natural fisheries if density-dependent mechanisms regulate abundance. Furthermore, depending on the homogeneity of aggregations of juveniles, individuals of nontarget species of commercial or ecological importance may be collected incidentally as by-catch, placing pressure on wild stocks of those species. For example, in the Philippines, milkfish fry constitute only 15% of collections and the 85% of by-catch is discarded (Primavera 2006). The effect of capture of wild fish as juveniles on target or nontarget wild populations will depend on the size- and age-structure of the population, population size, age at maturity, natural and fishing mortality rates, and quality and avail-ability of suitable habitat. Selective collection of wild juveniles can also have a negative effect on fish community structure, depending on the trophic role of the species and the complexity of the community structure. Depending on the gear employed, juvenile col-lection can result in physical disturbance of the substrate, impacts on benthic communities, and in the worst case, destruction of habitat. Finally, wild seed are undesirable from a biosecurity standpoint because the health status of wild-caught juveniles is questionable. Shrimp producers now perceive that postlarvae produced in a hatchery are less susceptible to important diseases than wild postlarvae.

In the culture of some species, sexually mature broodstock are collected from the wild and induced to spawn in hatcheries. Broodstock collection can also place pressure on wild fish resources. For example, in Thailand, Vietnam, and some other Asian countries, the availability of wild-caught black tiger shrimp (*Penaeus monodon*) for broodstock has declined drastically in recent years. Adult *P. monodon* broodstock are now rare in trawl catches. Capture of mature adults for use in shrimp hatcheries likely has been one of the reasons for the decline. This decline in *P. monodon* broodstock has been a major reason why shrimp producers in Thailand and other Asian nations have imported non-native, farm-reared broodstock of the Pacifi c white shrimp (*Litopenaeus vannamei*). This species has replaced much of the *P. monodon* culture in Thailand.

Predator Control and Other Effects on Wildlife

The concentration of potential food resources in aquaculture facilities provides attractive foraging opportunities for birds and other predators of cultured animals, which are viewed as a nuisance by aquaculture producers. The construction of aquaculture ponds has created new aquatic habitat while suitable habitat for wildlife with an aquatic orientation, espe-cially migratory waterfowl, has decreased elsewhere. Ponds are attractive to waterfowl and migratory birds as sources of food and shelter. Aquaculture structures such as net pens and shellfish rafts create a complex, three-dimensional environment in the water column that attracts and aggregates wild fish, their predators, birds, turtles, and marine mammals.

Aquaculture producers use a range of nonlethal and lethal techniques to control preda-tion of cultured animals. Most methods are intended to harass wildlife and induce them to move elsewhere. Although lethal controls are used occasionally, it is unlikely that lethal control will have negative effects on wildlife populations, especially of those for which aquaculture has allowed an expansion of population abundance. Most wildlife is protected by international conventions and national regulations. It is possible that location in migra-tory pathways and near breeding grounds could cause disruption to wildlife populations, but it is unlikely that these sites would be selected or permitted for aquaculture.

There is some risk that wildlife may become entangled or trapped in nets, ropes, and fl oating or submerged structures for fish and shellfish. This could lead to drowning or starvation, but properly designed, constructed, and managed facilities do not present a major risk of entrapment or entanglement. In sum, there are few negative interactions with wildlife and the probability of negative consequences is very low.

FROM ENVIRONMENTAL IMPACTS TO BETTER MANAGEMENT PRACTICES

This chapter provided a brief summary of the state of global aquaculture and the structure of aquaculture within the United States. Most of the chapter described various aspects of the relationship between aquaculture and the environments in which it is embedded. The focus has been on resource use and the specific environmental impacts associated with aquaculture production. The remainder of this book will endeavor to provide farm-level solutions to potential adverse environmental impacts of aquaculture. Throughout this chapter, we tried to emphasize that the impacts of aquaculture have broad spatial and temporal scales and, correspondingly, that solutions have wide-ranging context—from simple changes in farm management to complex solutions grounded in societal values. The better management practices provided in this book are intended to address near-field effects at a local level to improve the environmental performance of aquaculture.

2

Role of Better Management Practices in Environmental Management

INTRODUCTION

Best management practices (BMPs) are increasingly regarded as meaningful goals in the overall reduction of on-farm and processing impacts and, by extension, cumulative impacts of agriculture. The adoption and encouragement of BMPs are often seen as an end itself in attempts to achieve continuous improvement in environmental performance.

The assumption is that if the "best" practices are in place, producers are doing all that they can to avoid the worst impacts of production. This approach, whether BMP or GAP (good aquaculture practice), tends to divide the world into two camps—best or worst, good or bad. In reality, the implementation of a specific BMP will result in a range of environ-mental performance among producers, with most falling somewhere in the middle (Fig. 2.1). Although there can only be one "best" practice, many are likely to be "better" than a range of others. For that reason, throughout this chapter and most of this book, BMP refers to *better management practices*.

If continuous improvement is the goal, BMPs, by definition, are transitory and merely a means to an end, not the end itself. Furthermore, today's best practice will be tomorrow's norm and will eventually become the worst practice, to be avoided at some point in the future. If incremental improvement is the goal, surely today's BMPs will give way to even better ones tomorrow. For these reasons the term *best management practice* is gradually giving way to better management practices.

Fortunately the abbreviation BMP applies to both. Although subtle, the difference signals a fundamental shift in thinking.

One shift is the recognition that no single BMP reduces a key impact equally, whether considering all producers in a specific country or all producers of a particular species globally. There is no one-size-fi ts-all solution. The most effective BMPs depend on, among other things, the species cultured, type and magnitude of impact, scale of production, resources (capital, labor, land) available to the producer, and the overall management system already in place.

Another fundamental issue is that BMPs are often used as proxies for performance. It is assumed by many that the mere adoption of a BMP will always yield an acceptable improvement in performance and that specific BMPs will have consistent and predictable individual or collective impacts. But this has never been demonstrated for a range of dif-ferent producers with different intensities or scales of production. Given the constantly evolving nature of individual practices and combinations of them in complex management systems, there is constant change in the practices employed—and there should be.

Governments and certifi cation programs often use BMPs as proxies because it is easier to verify whether a practice is in use than whether an impact has been meaningfully reduced. Meaningful proxies, however, should be created from data and not be a substitute for it. In theory, BMPs are the easy way out—adopt this or that practice and everything will be all right. BMPs are easier—it is easier to tell people what to do than to go out on a site and determine whether they did it. In addition, many producers want to know what to do and don't want to have to think about it. This makes their life easier because they can comply much easier. In practice, however, BMPs are an inexact shortcut. How to think about achieving a performance result is more complicated.

Finally, the focus on BMPs tends to be prescriptive: if effects are this, then do this. Prescriptive solutions tend to result in compliance, but also may not. The achievement of meaningful results depends on whether the BMP can significantly reduce a particular impact and whether the BMP selected was the correct one and implemented properly by the producer.

Trends indicate a future with more targeted and measurable results and less attention to how results are achieved. Once an explicit part of regulations and standards, whether required or voluntary, BMPs are now relegated to guidance documents and workbooks. Thus, the current focus on BMPs globally is shifting from prescription to helping produc-ers improve performance against a baseline and giving producers the fl exibility to develop their own ways to improve performance.

This chapter explores the opportunities and limitations of BMPs as a tool for environ-mental management, improving overall environmental performance, and reducing waste in a world where reduced impacts are demanded by governments, investors, buyers, and nongovernmental organizations (NGOs). BMPs are also a tool for producers to increase their competitiveness in markets that are becoming increasingly competitive.

HISTORY OF THE USE OF BMPS

The desire to produce more, reduce effort, and reduce inputs and costs has probably been around since the dawn of agriculture. This is the impetus to fi nd better ways to undertake tasks such as producing food, fuel, building materials, and fi ber. Even domestication can be seen as a form of BMP, and it has been an ongoing process for each of thousands of different crops for thousands of years in many separate locations. In association with domestication, individual producers have been trying to produce more with less for nearly as long. Most of the history of BMPs represents an effort to reduce costs and increase production efficiency, whether in forestry, agriculture, aquaculture, or other industries.

The vast majority of BMPs have been invented by producers—not researchers, regula-tors, government extension agents, buyers, investors, or NGOs. In as much as there are better and worse ways to undertake virtually every known production activity, it goes without saying that BMPs exist for virtually every production activity.

BMPs are most easily promoted when they increase income immediately or at least in the short term. As market competition increases, BMPs can be promoted because they reduce costs and input use or increase overall efficiency. Finally, if producers are fi ned for waste or pollution, BMPs can also be used to reduce fi nes, liability, or risk. Traditionally, BMPs have been most easily disseminated when there has been a direct impact on income. BMPs that reduce impacts but do not have a positive impact on income are much less acceptable to producers. Yet, this can be the case when BMPs are proposed or required by governments or the private sector.

In the past few decades BMPs have been seen as a means to improve overall environ-mental management and performance. Nonetheless, there is considerable overlap between the two aims—reducing inputs and increasing efficiency of resource use per unit of pro-duction generally improves overall environmental performance. Among other things, ineffi cient resource use can result in pollution.

Forestry

BMPs have long been adopted in forestry as a way to increase production and production efficiency, reduce costs, and increase net profits. BMPs have been used to improve a wide range of impacts related to harvesting efficiency, reforestation and planting, forest growth rates, use of inputs, and processing.

Improved practices allow timber harvesting and conversion to be more effi cient. Clay and Amaral (2001) reviewed the potential of BMPs to improve the efficiency of conven-tional logging practices in the Amazon River basin while reducing overall environmental impacts. Nearly 20% of the logs cut in the Amazon are lost because cutters do not coor-dinate with skidder operators. Skidder operators, in turn, compact the soil and damage twice as many trees (including commercially valuable ones) when they do not map their evacuation routes. In turn, trails are cleared where there are no logs, and trails to access logs are often not the ones needed to pull them out. Planned logging roads are 33% nar-rower than unplanned ones. As a result they are cheaper to build and operate while having fewer environmental impacts.

Improved planning yields significant environmental gains in many cases. Log depots are not strategically placed in many cases, and unplanned operations use 2.5 times as much land for depots as planned operations. About 7% of tropical timber in Brazil is damaged by insects and rain before it is sawn at the mill. In addition to lost timber and needless environmental damage, operators also waste time and fuel. For every 1 m^3 of timber har-vested in unplanned operations, 2 m^3 was damaged, 73 km of forest roads were built, and 221 m^2 of canopy were opened (twice that of planned operations). Planned timber harvests extract an average of 25% fewer trees than unplanned operations. The second cut in managed forests yields more than twice as much commercial timber as in unmanaged forests (Clay and Amaral 2001; Clay 2004a).

BMPs can also be supported by governments to improve performance at a national level. In the 1960s, there was increasing concern about the ability of the United States to meet its own sawn wood requirements. A small team was organized to visit each of the thou-sands of sawmills operating throughout the United States. Within a decade the efficiency of sawn timber per cubic meter of roundwood was increased 40 to 60% (Clay 2004a).

Not only can technically focused BMPs lead to increased income and efficiency and reduced environmental impact, but social BMPs can also improve income and reduce impacts. Worker incentive programs increase overall productivity, reduce costs, increase net profits, and improve workers' income and job satisfaction. To date much less attention has been focused on this issue. However, such incentives will become an even more important factor as ever-increasing amounts of wood and pulp are produced in plantation forests and out-grower schemes.

Agriculture

In agriculture, as economic competition increases globally, the most effi cient producers will be the ones who survive. They will be defi ned by their ability to invent, identify, or adapt practices that reduce input use as well as waste and pollution. Such producers will be more profitable, or at the very least will remain competitive in the face of globally declining real prices. Those who remain competitive will not merely focus on how to produce a single commodity better. They will focus on their overall production systems and periodically evaluate what crops to produce to best utilize physical, fi nancial, and market advantages. The specific crops, like the practices to produce them, will change. This would tend to favor those BMPs that offer the best fi nancial returns rather than those that perhaps reduce key impacts most significantly.

Experiences in agriculture suggest that there are BMPs for larger and smaller producers as well as for those with more money to invest and those with less. As with other produc-tion systems, BMPs for agriculture vary by site, crop, scale and intensity of production as well as the human and fi nancial resources available to the producer. As such no single BMP or suite of BMPs will work equally effectively for all producers. In fact, any recom-mended BMPs are likely to favor some producers over others.

Because of their importance to the overall reduction of impacts and the sustainability of agriculture, the adoption of some BMPs cannot be left to the market alone. Most pro-ducers in the world will probably not make the transition to those better practices without support. Targeted government subsidies can, in the short term, provide incentives for the adoption of those BMPs that address impacts that are not affected by improvements in resource use efficiency. Government regulatory and permitting systems can also encourage the identifi cation and adoption of BMPs.

ROLE OF BMPS IN AQUACULTURE

The role of BMPs in aquaculture has been similar to that in other industries. They have been promoted by governments, development agencies, input suppliers, industry groups, academics, and NGOs as a way to reduce costs and waste, increase income, reduce pol-lution, produce higher quality products, gain or maintain access to new markets, and obtain regulatory relief. The assumption is that BMPs produce positive, verifi able, and predict-able results on the ground. The value of BMPs in measurably reducing key impacts is rarely monitored and therefore is much less clear than whether they have been adopted. Most BMP-based programs measure compliance (e.g., was a practice adopted) rather than performance.

The adoption of BMPs has been most rapid where there is a perceived benefit by the producer (particularly but not always with respect to income), where BMPs are a condition of permitting and licensing, or where adoption has allowed producers to avoid regulations. When BMPs are required by law, however, they may or may not always make fi nancial sense for a range of different-performing producers. There will be some need to identify the specific types of producers, performance levels, or cumulative impacts that may not be addressed by required BMPs as compared to the most effective BMPs.

BMPs are also adopted more quickly if they are seen to enhance a producer's "license to operate" (that is, if they are accepted by neighbors and others locally). In fact, social acceptance may actually slow the adoption of BMPs if they go against what is believed locally. For example, it was assumed for some time that building shrimp ponds in former mangroves was a best practice because that was where shrimp were found in the wild and that constructing ponds above the high-tide mark incurred unnecessary expenses. Similarly, retailers in the United Kingdom believed that salmon should be fed high concentrations of fishmeal and oil because that is what they would eat in the wild. These beliefs impeded adoption of better practices because they were based on faulty reasoning.

Reducing Key Impacts

It is difficult to reduce the key impacts of aquaculture before there is agreement on what the impacts are locally, much less globally. Given that considerable effort and expense may be required to reduce impacts, it will be very important to identify and rank impacts by their overall signifi cance. Although it is not easy to compare apples and oranges, if the number of impacts cannot be limited, it is likely that very little will be accomplished. Yet, if six to eight key impacts can be reduced through concerted effort and focus, it will be possible in the future to address the next tier of significant impacts.

BMPs are a tool—a means to an end—that can be used by producers to reduce key impacts. In the process of reducing key impacts, many BMPs also increase producer effi -ciency, reduce waste, and improve income. In many cases, BMPs are the ultimate win-win tool. This is precisely why producers are often keen to adopt them (when they know and understand the practices and any risks associated with them), buyers and banks want to encourage them, regulators want to require them, and NGOs encourage them in most of their industry engagement strategies.

Increasing Effi ciency

BMPs can help producers achieve increased efficiency. This translates directly into reduced use and costs of inputs. Less water exchange means less pumping, less energy use, and lower costs of production. Improved feed conversion means less feed used, better water quality, less pollution, higher profits, improved relations with neighbors and other resource users, fewer court cases and legal battles, and a more secure license to operate. Reduced disease means more stocked animals are harvested, higher feed conversion, less downtime, less quarantine, fewer conflicts with neighbors, and higher profits. Fewer escapes means less money wasted on animals that are never sold, higher feed conversion, fewer conflicts with others, and higher profits.

In short, any BMP that improves performance will tend to reduce some if not all other associated risks. There are, however, some key areas where the total benefits from all BMPs are compounded, i.e., greater than the sum of the parts. Perhaps more importantly, producers can select different BMPs that will allow each to achieve a comparable perfor-mance level (Fig. 2.2). Figure 2.2 shows the performance curves for three practices where some of the producers using each practice perform above the acceptable level, but not all producers using any single practice perform above the acceptable performance level. Thus, in the hands of different producers, each BMP gives a different performance and there is overlapping performance among BMPs. For some BMPs, most producers achieve acceptable performance; for other BMPs, only exceptional producers achieve acceptable performance. Producers may select different practices to manage water and water quality, feeds and feeding, diseases, and escapes. In doing so, producers will achieve different environmental performance levels with respect to effl uent pollution, invasive species, disease transmission, and genetic pollution.

Reducing Waste

Waste is currently defi ned as that which is not used effi ciently. Most improvements in overall efficiency will also reduce waste.

Aquaculture facilities can either produce physi-cal wastes or operate ineffi ciently (e.g., less biomass of product per unit of energy input). One of the greatest incentives for reduced waste and pollution is cost rather than regula-tion. From a producer's perspective, cost is probably the most important issue. In business, perspectives on waste have been largely influenced by input costs. The more expensive an input, the more wasteful it is not to use it effi ciently.

From a societal perspective, it may be important to defi ne performance levels that are acceptable and those that are not. Voluntary standards do this through market mechanisms. They are only voluntary, however, to the degree that producers want to be able to sell into different markets. Governments regulate and set additional requirements through permits and licenses.

In defi ning these performance levels, awareness of scarcity is a key factor in assigning value to an input. There is a general and increasingly reinforced perception that natural resources are finite, but the focus to date has been primarily on the environment and those natural resources that are perceived to have value. Wasting labor and employee's skills are not something most producers think about. Biodiversity, natural habitat, and ecosystem services are not perceived as having market value, even if they have intrinsic value. This perception is changing, and as it changes it will considerably influence the definition of waste and inefficiency.

Waste or inefficiency is a target that moves as a result of changes in science, public perceptions, market pressures, or a combination of all three. Concerns about water, carbon, energy, and greenhouse gases will very likely drive standards development in the future. The World Wildlife Fund, for example, is addressing such issues in aquaculture dialogues and standards development efforts. Tesco, the largest retailer in the United Kingdom, now requires all suppliers to calculate "food miles" to understand the carbon emissions associated with transporting food products to market.

What Has Been Learned?

Any use of natural resources has impacts. The questions, surely, are to identify the most important impacts and to defi ne acceptable levels of impact. By defi ning some impacts as more important than others, the magnitude or scale of the less important impacts are implied to be more acceptable or at least do not require initial or immediate attention. Besides, it is also clear that no one—producers, governments, buyers, or NGOs—can possibly work effectively on all impacts. To be effective, we must be more strategic.

By the same token, improved profitability and farm-level economic performance alone are insufficient as incentives to reduce every key impact of aquaculture. Given a choice, producers will selectively adopt those BMPs that improve net profits or reduce a significant impact without reducing overall income levels. Although reduction of key impacts over-laps somewhat with saving producers money or actually making them more, it is clear that some key impacts will not be reduced through market forces alone. The only way to know for sure is to identify the key impacts, undertake business-case (sensitivity) analyses for the BMPs and determine the fi nancial implications of various practices.

Incorporation of BMPs into Certifi cation and Ecolabel Programs

Most certifi cation and ecolabel programs claim that those producers who receive certifi ca-tion are actually best-in-class producers. They imply that certifi ed producers are better than average producers (Fig. 2.3) and that they should receive price premiums. But given that performance against a baseline for key impacts is only rarely, if ever, measured in any ecolabel or certifi cation program, this is more talk than action. Without metrics, com-parisons of ecolabel and certifi cation programs become subjective rather than science-based. Another issue is that some programs urge continuous improvement. Yet, without performance benchmarks (e.g., local, national, or global), it is impossible to determine whether someone is truly best in class.

Old Focus: Proscriptive and Practice-Based

Most certifi cation and ecolabel programs have focused on identifying what producers should and should not do, irrespective of whether the focus has been on better management practices, best management practices, or good aquaculture practices. Adoption of a single practice has been assumed to be the best way to improve overall performance across the board, but it is not. It is convenient, easier, and probably cheaper than the alternatives, but no single practice will predictably and uniformly reduce a key impact across all producers.

Figure 2.2 shows the problem, which is in fact two problems. First, given the wide range of producer realities, no single set of practices will reduce key impacts to acceptable levels. Second, producers who adopt a particular practice will have widely different on-the-ground results. One size does not fi t all. This is true when addressing a single impact, more than one impact, or cumulative impacts. In short, whether BMPs work or not will be hit or miss. The point here is that BMPs do not guarantee performance. If a performance result is what is wanted, it should be asked for and the producer should be allowed to fi nd the best way to achieve it.

If the goal is to reward better producers, the certifi cation program needs to be able to single out such producers based on performance, not practices. In many cases, BMPs are not entirely or even mostly transferable. What works for one will often not work for another. Furthermore, good performance in one area may cause poorer performance in another. For example, high fishmeal, high-fat diets produce better feed conversion (less waste) in salmonids, but the trade-off is the increased use of fishmeal. In shrimp produc-tion, water exchange improves water quality but uses tremendous amounts of nonrenew-able energy.

The issue is about balanced thinking. More specifically it is about *how* to think, not *what* to think. Proscriptive programs yield compliance that may or may not yield the desired results. If the required practices are well targeted, such programs may result in continuous improvement but not necessarily in acceptable performance levels. Continuous improvement for very poor practices might in 10 years result in a producer achieving the industry norm. Unfortunately, by that point the norm will have improved. For example, Chilean salmon farmers used significantly more chemical therapeutants per tonne of pro-duction than their counterparts in Norway a decade ago. Over the past decade, Chilean farmers have improved their performance, but have lost even more ground to the Norwegians who have done an even better job of reducing the use of therapeutants. There is no acceptable performance level that is written in stone. What is acceptable is based on what is possible—and that is constantly improving. By their very nature, what is certain, however, is that proscriptive-based BMP programs will encourage compliance, not innovation.

New Focus: Performance-Based Standards or Results

Increasingly there is awareness that ecolabels and certifi cation programs imply improved performance standards that cannot be or are not verifi ed on the ground. This is the case because few, if any, ecolabel or certifi cation programs actually measure impacts, either before or after producers join the program. Ideally, such programs would incorporate global or regional performance benchmarks and then be able to show credibly how their program ensures that producers exceed those standards.

Thus, BMP- or GAP-based programs assume better performance but cannot guarantee it. Continuous improvement, although important in some respects, requires benchmarking if it is to be credible. There is nothing to prevent a poor performer from continually improving for years and years and still be a far poorer performer than others who are better in class.

Performance-based standards quickly raise and give direction to other important issues for discussion. What is the appropriate scale for performance: A pond or net pen? A facil-ity or business unit? A company? An ecoregion? Consideration of scale also allows clear definition of a significant impact. For example, is it energy use per tonne of production in a pond or net pen, food miles to market, the energy that is embedded in feed that is used to grow or catch feed ingredients, transport ingredients and feed, etc.? Clearly it depends on the regulator, the buyer, the certifi cation program, or more likely than not, awareness at the time. For example, most people do not realize that most of the embedded energy in aquaculture-produced salmon is in the fishmeal and oil in the feed, not in the production system or the transport of fish to market. So, once again, the question is what are the key impacts and what is the best scale to look at them? While reasonable people may disagree on these issues and while some may be too cumbersome or expensive to measure, a focus on practices does not lend itself to this kind of approach.

Within the context of performance-based standards, BMPs are means to various ends. However, it is important to recognize that they are not ends themselves. Thus, BMPs do not belong in standards; they belong in workbooks and guidance documents. Such documents are more widely useful when there are several BMPs that can be shown to achieve desired performance levels, depending on the specific resources and conditions of a producer. Figure 2.2 demonstrates how performance curves for different practices would allow individual producers to meet overall performance requirements of any given program. Of course, to the extent that guidance documents can also provide concrete information on the investments required, the return on the investments to adopt BMPs, and the overall payback period, the more likely and quickly others will adopt them. Finally, it is important that such documents identify the BMPs that are known to favorably complement and reinforce other practices. Those BMPs that address one environmental impact, but inadvertently create another undesirable impact must also be identifi ed.

ROLE OF BMPS FOR DIFFERENT GROUPS

As BMPs have been developed, adapted, and adopted by producers, many different groups have seen their value as proxies. Producers see them as a way to make money. Buyers see them as an indication of higher-quality product that is associated with fewer risks. Investors see them as a way to differentiate between better and worse credit risks. Regulators use them to defi ne the line between good and bad performance. The role of BMPs for each group involved with production is discussed below.

Producers

Producers invent most BMPs, often in response to a specific problem (e.g., disease, increased costs of inputs, regulations, market requirements, etc.). In the past, BMPs were seen as tools for individual producers to adopt or not. Increasingly, however, producer groups are beginning to realize that the reputation of all can be affected by a few. This can affect local reputations, market access, or government regulation. As a consequence, BMPs are now an important part of producer codes of conduct. Groups of salmon, shrimp, catfish, shellfish, and trout producers—to name but a few—have all identifi ed, agreed to, and incorporated BMPs into codes of conduct and certifi cation programs.

Many producers are reluctant to share fi nancial information because they fear it will be used against them or by their competitors to gain market advantage. Most producers are interested in increasing production and/or net profits and any funding they have would be directed at those issues. Most governments, by contrast, tend to guide what resources they have toward identifying BMPs that contribute significantly to the reduction of targeted impacts.

Research Scientists

Although the vast majority of BMPs (the total may be 90% or more) were originally invented or discovered by producers, it is not the role of producers to document or explain how BMPs work, how they can be adapted and disseminated for different growers, or the cost of the BMP and the fi nancial payback for different types of producers. This work falls squarely in the domain of research and extension scientists. At this time, however, there is too little funding to undertake this work. Many of those who know and can describe the technical aspects of BMPs do not have the skills to understand and describe the fi nancial implications.

Seafood Buyers

Seafood buyers are increasingly using BMPs to discourage worse production practices or encourage better ones. Buyers are motivated by a desire to ensure basic seafood health and safety and other product quality issues, while simultaneously avoiding reputational risk issues that could be associated with products produced in ways that can be linked directly to harmful impacts, particularly ones that are of increasing public awareness and concern.

Programs like EurepGAP require that producers comply with the legal requirements of the country where the product will be sold. The emphasis here is clearly on health and safety and food quality. Although producers are usually required to meet European Union standards to sell into those markets, those standards are often significantly higher than those that many producers are required to meet in their developing country homes. In some contexts, the practices used to meet local or international standards are GAPs. This con-fuses the issue somewhat as it implies that anything that is legal is good and anything that is not is bad. Although the latter may be true, the former probably is not.

The Global Aquaculture Alliance/Aquaculture Certifi cation Council program has a similar approach. However, in this case, although producer certifi cation standards exist, some 90% of certifi cations to date have been for processing plants, demonstrating the clear preference for food quality and health and safety concerns fi rst. Although this is clearly the preference of GAA supporters like Darden Restaurants, this approach does not address the key on-farm social and environmental impacts of shrimp aquaculture.

Other buyers want the products they sell to be produced to higher standards. They achieve this by purchasing products from producers on the most effi cient or best end of the performance curve for key issues. Buyers who seek product from producers who are best-in-class now purchase from those who are certifi ed by third parties and ecolabel programs, or work with NGOs or independent consultants to create their own standards. Either way, the goal is to go well beyond compliance with either local or importing-country laws by purchasing from producers who have adopted or often created the best-known practices to reduce each of several different impacts. Whether they can actually do this without measurement is another question.

Buyers are rarely interested in paying a premium for product, whether or not the premium encourages better producers or discourages worse ones. They may enter into longer term contracts to buy products, but they will rarely lock in price ahead of time or for a longer period of time (e.g., 6 to 12 months). Buyers may pay a higher price for a product with specific attributes if demand for such products increases faster than supply. This means that producer-level BMPs need to either pay for themselves directly or, if undertaken together, have a net positive impact on income.

Because buyers will generally not pay more for products, any certifi cation or verifi cation system that ensures that BMPs have been adopted must be focused, limited in scope, and inexpensive to undertake. Also, to the extent that buyers care about specific issues, they will want to verify producer claims of reduced impacts against a baseline. All of these reasons suggest that BMP-based certifi cation or verifi cation programs should be limited in scope. Fortunately, much of the research suggests that only six to eight activities cause most of the impacts that the public cares most about and that might affect buyer reputations.

Investors

Given a choice, private-sector investors tend to evaluate projects based solely on fi nancial risks and returns. In the past, most of the focus has been on the fi nancial viability of the entity borrowing money. Over the past 30 to 40 years, governments and multilateral lending agencies (e.g., the World Bank, the Asian Development Bank, and the International Finance Corporation) have incorporated social and environmental impact assessments into project evaluation, increasingly making it a condition of lending for aquaculture projects. Over time, some private sector lenders have seen that reduced environmental and social impacts actually implies better overall management, which has positive fi nancial implications.

Regulators

Regulators are increasingly interested in fi nding ways to use BMPs to improve perfor-mance and reduce waste and pollution. For many, BMPs are a proxy for achieving regu-latory compliance and/or a condition of permitting or licensing. However, regulators face the same problems as other nonproducers who try to use BMPs as a proxy for performance: each BMP will yield a range of results. In fact, most regulators, at least in the short term, want to eliminate the worst impacts (Fig. 2.4). The thinking is that those who produce those impacts must be using worse practices. Although this may be true to some extent, it certainly is not true across the board. Any number of practices can yield unacceptable impacts, but it is quite likely that many of those same practices in the hands of other producers could have acceptable results.

This caveat aside, there may be times when a BMP-based approach works for regula-tors. This would most likely occur where production systems are similar throughout a regulated aquaculture industry. In this case, it is more likely that BMP-based regulatory approaches will yield expected and desired results. This is perhaps true for more mature industries or in regions where producers have adopted the same technological systems for the production of the same species.

The variation in appropriate and effective BMPs by species is an added complication for a BMP-based regulatory system. What works for one species will not work for another. There are even differences among the BMPs for different bivalve mollusks (e.g., clams, oysters, mussels, and scallops) in terms of how they are produced and harvested. In effect, this means that regulators need to have robust BMP-based regulatory systems for each of the main species or species clusters if they are going to have the desired impacts and refl ect species-specific differences.

Finally, as mentioned earlier, BMPs change—in some cases relatively rapidly. Regulators need to revisit any BMP-based producer requirements regularly (e.g., every 3 to 5 years). This is further complicated by the considerable lag time that often exists between when BMPs are fi rst invented and when the consensus is built to the point that they become acceptable as a basis for regulations, permitting, or licensing.

Another consideration is whether a BMP-based regulatory system actually encourages or discourages the identifi cation of new BMPs that can reduce the key impacts even more. BMPs are usually invented in order to solve a problem. In many cases the problem is perceived as regulatory compliance. Government regulators ask producers to do something (e.g., reduce risk, waste, or pollution) that may be beyond what many are doing currently. Regulators telling producers what they can do to allow them to be in compliance (i.e., adopt a BMP) will not really encourage innovation. Many BMPs are invented in other countries or for other species. A BMP-based regulatory system may not sufficiently encourage producers to go beyond compliance in their efforts.

In the long term, shifting the performance curve of most producers against previous benchmarked performance will allow government to develop stricter regulatory, permitting, or licensing systems. In less than 10 years, performance levels that are deemed unacceptable can be as high as what was previously considered to be above-average performance in any given system (Fig. 2.5). This shift is possible because of innovation, the creation of a new set of BMPs, or because a voluntary program set a high-performance goal that challenged industry to perform at a higher level. Although it is not clear that regulators can ever set the performance bar so high, it is clear that society can and should take advantage of new norms that are shown to be profitable in voluntary programs.

LIMITATIONS OF BMPS

In addition to the obvious limitations of an approach that focuses on processes and prac-tices rather than measurable performance levels, BMPs as currently defi ned have other limitations. They do not allow producers—individually or collectively—to understand, much less reduce to acceptable performance levels, their cumulative social or environ-mental impacts. BMPs for a single producer operating in relative isolation will be based on what is fi nancially and technically feasible. However, acceptable performance levels for many producers operating in the same area may be much more stringent because of the cumulative impact (e.g., water use, effl uent quality, etc.) associated with a concentra-tion of production. Most BMP-based systems are not focused on and generally do not address social issues at the producer level, so they will have no ability to predict com-munity or regional issues. As mentioned before, specific BMPs or packages of BMPs will inevitably favor some producers over others. By focusing on existing BMP adoption rather than results, there may be little incentive to develop or invent new practices or to look to other industries for insights as one would have to do with increased performance require-ments. Although it is beyond the scope of this chapter to do more than speculate, there may be a lag time for ecological impacts that may be either simply too long to measure or too late to change once identifi ed. This suggests that BMPs are means to various ends. However, they will not spur innovation or improve performance over time. In fact, the best way to address cumulative impacts remains an open question.

Cumulative Environmental Impacts at a Landscape or Ecoregional Level

Virtually all BMPs focus on what producers can do individually to address identifi ed impacts. However, from an environmental perspective, the impacts of a single producer using BMPs will not produce results that are significant at the landscape or ecoregional level unless that producer is quite large or the impacts particularly egregious. At some point beyond the individual producer, however, small changes made by many producers can make a difference.

The carrying capacity of a landscape or ecoregion can be exceeded even though indi-vidual producers have adopted and implemented BMPs. The point here is that even with BMPs, there is a limit to how much activity can take place before the cumulative effect exceeds the capacity of the surrounding area (defi ned at an appropriate scale) to assimilate that impact. Different BMPs may allow more production, others less. However, aggregate impacts are not limited to the number of producers or widespread adoption of a specific BMP. Carrying capacity takes all practices into account. Strategically, it would be best to focus on those practices that cause the most significant impacts individually or whose impacts are compounded when combined with other practices.

Finally all the BMPs in the world will not correct the problems that can be caused by siting aquaculture production facilities in the wrong place (Clay 2004b). This may be due to impacts on critical habitat, biodiversity, or environmental services. Some locations may become obsolete because technology has changed or our understanding of the most sig-nifi cant impacts has changed. For example, through the early 1990s, shrimp farmers were told by technical consultants that it was best to construct ponds in former mangrove areas. They were also told to use 20% water exchange to maintain water quality. Now we know that neither of these is necessary.

Increasing awareness of impacts can also suggest not only new issues, but also perfor-mance standards and the BMPs that might help to achieve them. A good example of this is an increasing awareness of how carbon, energy, food miles, and climate change are affected by different production systems and practices. Tesco has asked all its suppliers to give the company a number for the amount of carbon embedded in the food miles for each product Tesco buys from them. Energy-intensive systems may be advisable for some producers, but they are now being looked at through a different, energy-effi cient lens. In the future, water, biodiversity, liability and risks for escapes, etc., are likely to be similar issues. Impacts are not static, because science and the awareness that it drives is not static.

Larger Social and Political Issues

Although BMPs can be developed to address any issue, they have typically been used to help workers reduce environmental impacts to more acceptable levels. Producers, for their part, see BMPs as a way to improve performance and increase efficiency and net income. Thus far, BMPs have not been widely seen as helpfully addressing social issues. Some producers have shown that incentive programs increase their net income by two to four times, and adult literacy programs can reduce worker turnover rates within 2 years from 80 to 20% per year and improve efficiency at the same time. Although there is a lot of work that could be done on BMPs that improve social performance, to date very little is known about their costs, their effects, and their impacts on profitability.

On the social side, the focus of ecolabel, certifi cation, and government regulatory pro-grams is limited mostly to obeying the law. Even certifi cation programs that promote fair trade are focused mostly on the prices paid to producers (and obeying the law), not the conditions of the labor hired in aquaculture production or processing facilities or the impact of operations on neighboring communities. Simply put, there has been little intel-lectual or political interest to address more broadly defi ned social issues within BMP programs. Enlightened producers have found ways to improve their social performance and many have been rewarded by increased profits as well as in a number of less tangible but locally important ways.

With the exception of those issues about which there is consensus and the issues that are already codifi ed in labor and other laws, there is little agreement about what other issues can or should be addressed by BMPs. In the nine aquaculture and agriculture com-modity dialogues that World Wildlife Fund is helping to convene, there is broad and generally easy consensus about environmental impacts and therefore the types of BMPs that might reduce them. On the other hand, there is little or no consensus about the key social impacts for any single species and the different ways it may be produced, and even less across a wide range of species and each with a range of production systems. In part this is cultural. Different countries have different social and cultural values and different ideas about what government, producers, and society as a whole should do. Where there is agreement, the BMPs tend to be codifi ed in law. However, there is often little consen-sus nationally beyond what is in the law, and little if any global agreement on these issues.

Many Are Best for Specifi c Types of Producers

BMPs tend to be linked to specific types of production systems or to specific product quality and handling issues. The party who requests or requires BMPs will have much to do with specifying BMPs they want to see. For example, buyers will want practices that enhance product quality and address known health and safety issues. Some producers will have a comparative advantage over others in complying. In addition to species cultured, the key variables appear to be scale, intensity, producer country norms, availability of capital, and cost of labor. The issue is not just that BMPs are not one-size-fi ts-all, but rather that there are BMPs that will make every producer perform better.

Many certifi cation and ecolabel programs, particularly those that intend to be global, are based primarily on the reality of a certain type or class of producer. Such programs can inadvertently exclude others. A BMP for extensive, semiextensive, intensive, or super-intensive production systems will vary considerably. Similarly, the impacts that are accept-able for black tiger prawns (*Penaeus monodon*) are different than those for Pacifi c white shrimp (*Litopenaeus vannamei*). Because the biological and cultural requirements differ for the two species, performance standards will differ for feed conversion, stocking densi-ties, survival rates, depth of the water, and harvest times. This is also true of basa (*Pangasius bocourti*) and channel catfish (*Ictalurus punctatus*) and trout or tilapia raised in different ways in different places.

From the perspective of a global program, it is easiest to have a single program with standards that are clear to the producer and easy to verify. It is also clear that many buyers, brands, and retailers would—all things being equal—prefer to support those BMPs that refl ect the types of systems that they think best address food quality and health and safety issues.

No Home for BMPs

Many producers do not have access to information about BMPs that would allow them to identify, much less adopt, those BMPs that might best serve their needs. In today's world of instant information it often takes up to 10 years between when a BMP is invented and when it is made more widely available to others. This diffusion time period is faster than other industries, which typically adopt a new technology at a large scale 50 to 70 years after invention (Grubler 1998). For BMPs, the inventor will test the practice for a few production cycles and then gradually roll it out. It will become known and talked about locally, and then a researcher from a university or a government agency will appear. They will spend time documenting what is going on, undertaking independent trials, and having their results reviewed and verifi ed by others, and eventually after a lag of a year or two they will write up the results and get them peer reviewed and then published.

By this point the producer who originated the BMP has invented something more or better that supersedes the last one. What is being rolled out globally is already obsolete. We need to shorten the amount of time that this process takes. To do that we will need to be more strategic. If we assume that we can look at the practice in isolation rather than document how it fi ts into the entire production system this will speed things up consider-ably. Also, because BMPs cannot be transferred indiscriminately from one farm to another, we can assume that what is important is how the inventor thought about the practice and how it fi t into the operation rather than documenting the specific details of the practice. Here again the issue is how to think, not what to think. This kind of information, and the general fi nancial and other implications, can be captured and conveyed far more quickly than has been done to date.

Even so, there is a need for a home or several homes for BMPs. It would be a tremen-dous boon to the aquaculture industry if BMPs, the results they achieve on the ground, and their impact on production and income could be documented and made more readily available to others. This can be accomplished through a dedicated website. Although many aquaculture producers would have access to such technology, many, if not most, around the world would not. Radio programs might be effective in some areas, farmer field schools like those run by the United Nations Food and Agriculture Organization (FAO) or farmer exchanges could fill part of the niche, and certifi cation programs could as well. Government extension programs would be ideal but they have less funding than in the past. Increasingly the private sector is doing this through contract producers, processors, input suppliers, or loans.

Whatever the delivery mechanisms, it would be the full-time job of several people to develop such information and deliver it to and through the appropriate channels. It would probably need to be in one language initially, but relevant parts (e.g., impacts, species, and practices) could be translated as appropriate.

Given the transitory nature of BMPs, such information would have to be updated regularly. However, the goal is help producers better understand how other producers from around the world have been able to solve problems that affect them. Clearly today's BMP is tomorrow's norm and the one that needs to be replaced the day after.

CONCLUSIONS

Better management practices are better than worse practices. No surprise there. They are also better than either good or bad aquaculture practices. They are not, however, a credible proxy for performance. If we are to strategically reduce the most significant impacts from aquaculture production, we must fi rst agree on what they are. Science-based consensus is very important on this issue, particularly when there is so much information of dubious value in the public domain and even in peer-reviewed articles.

Once we have agreed on the key impacts of different aquaculture production systems, the task is to measure those impacts at this point in time. If we cannot defi ne the range of performance for a given impact, it will be hard to draw a line regarding what is acceptable and what needs improvement. Because aquaculture is a global industry and products traded around the world compete with each other, the goal should be to understand average global performance benchmarks and why and how they vary around the world.

There is an old adage: If you don't know where you're going any road will get you there. This is only half true. It is essential to know where we are *and* where we are trying to go. Reasonable people will disagree about priorities and strategies under the best of conditions. Knowing where we are globally and what the range of performance is will be essential to developing a viable strategy for moving forward. There is considerable evi-dence to suggest that with regard to the same potential issue, some producers can have 10 to 100 times more impact than others. When our thinking about goals or current realities is based on assumptions, impressions, or concerns rather than science and reliable data, the disagreements will be even more intense. Credibility comes with metrics.

As important, innovation comes from metrics. Innovation comes from clearly agreeing on and defi ning a problem and letting each producer fi nd the best way to achieve accept-able performance levels. Setting challenging performance standards, not identifying a set of BMPs, will be key to the future of sustainable aquaculture, because as soon as a BMP is written in stone, it is obsolete.

3

Better Management Practices in International Aquaculture

INTRODUCTION

The United States imports about 85% of its fisheries products, and fish, shrimp, and mol-luscan shellfish originating from aquaculture are a major component of this trade. Several environmental nongovernmental organizations (NGOs) became highly critical of aquacul-ture and particularly about marine shrimp and salmon farming in the 1990s. Attempts by more radical environmental NGOs to effect consumer boycotts of these species in the United States and other western nations failed. However, several moderate NGOs persisted in efforts to promote environmental and social responsibility in aquaculture. Criticism of aquaculture practices by World Wildlife Fund, Environmental Defense, Conservation International, and other environmental NGOs has caused aquaculture organizations to promote environmental responsibility, governments to increase environmental regulations for aquaculture, and some retailers and other food service buyers to seek environmentally and socially responsible aquaculture products.

Most negative environmental and social impacts of aquaculture occur at the farm level, and particular emphasis has been given to encouraging producers to adopt best management practices (BMPs) designed to lessen these impacts. The NGOs also have succeeded in obtaining the participation of the United Nations Food and Agricultural Organization (FAO), World Bank, and other international development agencies in efforts to promote responsible aquaculture. Moreover, educa-tional programs by NGOs to encourage more responsible seafood choices by consumers apparently are beginning to influence the market.

Aquaculture in the United States has been subject to various federal and state environ-mental regulations for years. However, the United States Environmental Protection Agency (USEPA) only recently developed a federal rule for aquaculture effl uents. This rule-making activity was initiated because of intervention by environmental NGOs. And, environmental NGOs appear to have become interested in United States aquaculture mainly because of their experiences with salmon and shrimp farming in other nations. USEPA decided that domestic aquaculture is not a major source of pollution. The effl uent rule basically recommends adoption of BMPs by operations large enough to be considered concentrated aquatic animal production facilities. Changes in aquaculture sourcing policies by major buyers seeking products from responsible aquaculture likely will have greater effects on United States aquaculture than the USEPA effl uent rule and other environmen-tal regulations.

The purpose of this chapter is to provide an overview of international aquaculture, discuss programs for improving environmental and social responsibility of interna-tional aquaculture, and consider possible effects of these programs on United States aquaculture.

WORLD AQUACULTURE PRODUCTION

World aquaculture has grown from a few million tonnes annually in the 1950s to nearly 60 million tonnes in 2004. Statistics from FAO (2006) are used below to provide an over-view of current world aquaculture production.

Production by Environments and Species Categories

Aquaculture is considered by FAO to be done in three environments: freshwater, brack-ishwater, and marine. In 2004, total aquacultural production for the three environments were 25.8 tonnes in freshwater, 3.4 tonnes in brackishwater, and 30.2 million tonnes in marine environments. The data are summarized by major species groups (Table 3.1). Freshwater aquaculture is largely fish culture, crustacean culture dominates brackishwater aquaculture, and aquatic plants (primarily seaweeds) and mollusks (mainly bivalves) are the main marine aquaculture crops. If only aquatic animals are considered, aquaculture production in freshwater is considerably greater than that of brackishwater and marine environments combined.

Further analysis of the FAO data (Table 3.2) reveals that carp species dominate fresh-water production, but there also is significant production of unspecifi ed fish species and tilapia. Production of Pacifi c white shrimp (*Litopenaeus vannamei*) and black tiger shrimp (*Penaeus monodon*) dominates brackishwater aquaculture. Bivalve mollusks are the main categories of marine animal culture, but Atlantic salmon (*Salmo salar*) is also significant. Of course, marine plant production is slightly greater than bivalve mollusk production in marine aquaculture.

Geography of Aquaculture

The world is divided into six aquaculture regions by FAO. The following are the regions and the 2004 production of each: Oceania, 139,273 tonnes; Africa, 570,113 tonnes; North America, 955,178 tonnes; South America, 1,137,825 tonnes; Europe, 2,238,683 tonnes; Asia, 54,367,372 tonnes. Asia provides 91.5% of world aquaculture production.

Although many nations have aquaculture, relatively few nations provide most of the world's production of cultured aquatic organisms. Nine of the top ten and twelve of the top 20 aquaculture nations are in Asia (Table 3.3). According to FAO (2006) statistics, China is responsible for more than 50% of world aquaculture production, with more than 41 million tonnes annual production. The FAO can offi cially report only fisheries data provided by governmental sources in each nation. Most experts agree that the Chinese government has traditionally infl ated their aquaculture data—possibly by a factor of 2 or 3 (Boyd et al. 2005). Nevertheless, even if the Chinese data are infl ated fi vefold, China still has more than twice the aquaculture production of India, the second largest aquacul-ture producer.

Aquaculture Products in International Trade

The top ten species and species groups produced by aquaculture (Table 3.4) include salmonids, tilapia, shrimp, oysters, clams, scallops, and mussels, popular with consumers in developed nations where a high percentage of seafood is imported. For example, about 2.1 million tonnes of seafood are consumed annually in the United States, of which about 80% is imported. Average per capita consumption of seafood in the United States is 7.4 kg, and 10 species or species groups account for over 90% of the total consumption (Table 3.5). Aquaculture is an important source of shrimp, salmon, catfish, crab, tilapia, clams, and scallops for United States consumers. Only one species listed in Table 3.5, channel catfish (*Ictalurus punctatus*), is almost totally from domestic aquaculture. Two species not listed in Table 3.5, red swamp crayfish (*Procambarus clarkii*), known in the United States as crawfish, and rainbow trout (*Oncorhynchus mykiss*), also are popular with United States consumers, and these two species are produced mainly in domestic aquaculture.

CONTRASTS BETWEEN UNITED STATES AND INTERNATIONAL AQUACULTURE

Environmental issues in aquaculture are described in Chapter 1. There are, however, major differences between aquaculture in the United States and other developed countries and aquaculture in developing nations. In developed nations, most seafood is imported from developing countries, and aquaculture is limited mainly to a few species of moderate or high value for domestic markets. For example, channel catfish, rainbow trout, and crawfish produced in aquaculture in the United States are almost exclusively marketed domestically. Several countries in the European Union produce trout, salmon, and bivalve shellfish by aquaculture. Although some of this production is exported, much is for the European market. Carp production in Europe is for domestic markets.

In developing nations with large aquaculture sectors, there is greater variety of culture species than in developed nations. Although most of the production is consumed domes-tically, some developing countries export large quantities of aquacultured shrimp, tilapia, salmon, and several other species.

Farms in developed countries usually must be fairly large to be profitable, and there are relatively few producers. For example, in 2006 there were 1,035 catfish farms in the United States and the average farm size was 67 ha (USDA-NASS 2007). Labor is expensive, so mechanization and automation is used as much as possible. Even family farms must invest in the latest technology to be profitable. There are tens of thousands of small aquaculture farms in many developing nations. In pond culture, farms typically have water surface areas of 0.25 to 5 ha. Other production systems also are much smaller than those in developed nations. There are some extremely large shrimp, tilapia, and salmon farms in developing nations, but the combined production by small farms in many nations, especially in Asia, greatly exceeds that of large farms.

Small-scale farmers tend to be poor and live in rural areas. They are poorly educated, lack knowledge about aquacultural technology, and are unaware of environmental issues. Production practices often are substandard, and few producers use BMPs to lessen nega-tive environmental impacts. Large farms usually hire highly qualifi ed managers and use better production practices than small farms. Because labor is inexpensive in developing countries, large farms are not as mechanized as those in developed countries. Nearly all large farms in developing countries focus on the export market.

There often are dense concentrations of small aquaculture farms in developing nations (Fig. 3.1). The environmental implications of such dense clusters of farms are obvious, and the likelihood of negative impacts often is exacerbated by lack of effective govern-mental regulation of effl uents, wetlands, predator control, water use, and other factors. The combined effect of many small farms on the environment can be as great or greater than that of a large farm.

Aquaculture facilities in developed nations also tend to be concentrated in specific areas. In the United States nearly all catfish farms are located in a few counties of Alabama, Arkansas, Louisiana, and Mississippi. In Alabama about 50% of catfish farms in the state are on the catchment of a single creek, but the farms occupy only 7.5% of the surface area of the catchment (Silapajarn and Boyd 2005). The largest portion of the United States trout industry is located along a stretch of the Snake River in Idaho. Nevertheless, the facilities usually are separated by distances of 1 to several km. Environmental regulations also are imposed on aquaculture in wealthy nations, and there are few alleged or docu-mented negative environmental impacts of aquaculture.

Developed nations generally have more equitable social conditions than found in devel-oping countries. In many poor countries, agricultural workers may not be paid a legal minimum wage, working and living conditions on farms may be inadequate, and there may be no benefits or job security. Large aquaculture companies may be given priority over other stakeholders for use of land, water, and other resources, resulting in conflicts with local communities. Companies may hire laborers from other places in preference to local workers. Child labor also may be used at some farms. Small-scale producers usually do not hire workers, but underaged family members are expected to work on farms. This is acceptable provided children are not denied the opportunity for an education. Obviously, social injustices are more likely to result from aquaculture in developing nations than in the United States or other developed nations.

Environmental NGOs continue to criticize aquaculture on environmental and social grounds, but discussions with offi cials in several of these organizations suggest a change in attitude. They have come to realize that aquaculture is a necessary component of the world's food production system, and that when done properly, aquaculture can be more environmentally responsible than fishing (Boyd et al. 2005). Fishing is analogous to hunting and gathering, while aquaculture is a form of agriculture. Humans no longer depend on hunting and gathering as a primary source of food and fi ber from terrestrial sources. Likewise, we should increasingly turn to responsible aquaculture as an alternative to fishing for sources of aquatic food.

CODES OF CONDUCT AND BMP PROGRAMS

There are a number of initiatives to improve the environmental and social performance of international aquaculture. These range from programs encouraging voluntary adoption of BMPs to certifi cation programs in which BMPs are implemented as a means of complying with verifi able environmental and social standards. There has been widespread use of BMPs to improve resource use and lessen the amount of nonpoint source pollution in traditional agriculture (Hairston et al. 1995; Clay 2004). Thus, the initial response of promoting codes of conduct and BMPs in response to the criticism of environmentalists over negative impacts of aquaculture was not surprising (Boyd 2003). The efforts mostly were initiatives of aquaculture associations, government fisheries and aquaculture agen-cies, and international organizations that support aquaculture development. It is notewor-thy that the promotion of better practices began well before there was concern in the international market about environmental and social issues related to aquaculture.

FAO Involvement

The FAO has a long history of promoting better aquaculture practices to increase produc-tion, improve efficiency of resource use, and avoid negative environmental and social impacts. The FAO Fisheries Department (1997) published a *Code of Conduct for Responsible Fisheries* that has served as guiding principles for many of the recent respon-sible aquaculture programs. Shrimp aquaculture has been a particular focus of FAO. Expert consultancies were held in Bangkok, Thailand (1997); Rome, Italy (1998); and Sydney, Australia (2000) to discuss the environmental impacts of shrimp farming and to develop best management practices for preventing these impacts.

The Consortium on Shrimp Farming and the Environment

The World Wildlife Fund (WWF), Network of Aquaculture Centres in Asia-Pacifi c (NACA), and World Bank joined with the FAO and the United Nations Environment Programme (UNEP) to form a consortium to investigate shrimp farming practices world-wide. The consortium conducted case studies of shrimp farming in several countries. For example, in Ecuador, studies were conducted on capture of wild postlarvae, farm effl uent composition, and effects of shrimp farming on mangroves. Thematic studies that cut across countries were prepared for topics such as chemical use, inland shrimp farming, shrimp diseases, etc. Many of the case studies have been posted at the NACA website (www. enaca.org). The fi ndings of the case studies were used to develop better management practices for shrimp farming (FAO et al. 2006).

Government Programs

Aquaculture products are important sources of foreign exchange in many countries. Thus, governments of nations with significant aquaculture have the dual task of protecting the environment and promoting the export of aquaculture products. Some countries have programs to assist producers in developing and adopting codes of conduct and BMPs for responsible aquaculture. The government of Madagascar developed a code of conduct for shrimp farming with recommended practices, including some legal aspects to be con-sidered later (Edaly 2003). The Thailand Department of Fisheries devoted much effort to promoting a code of conduct (CoC) program for marine shrimp farming (Tookwinas et al. 2000). Recently, the Thailand Department of Fisheries also initiated a good aquaculture practices (GAP) program for Thai shrimp farming that is less strict than the CoC program.

Producers can choose between the CoC and GAP programs (Waraporn Prompoj, Thailand Department of Fisheries, personal communication). The Malaysia Department of Fisheries also made a CoC for shrimp farming that is an example of general guidelines for voluntary adoption by producers.

Industry Efforts

Aquaculture industry associations in several nations have developed environmental policies, codes of conduct, or codes of practices for their members. One of the fi rst and best-known programs is the code of practices of the Irish Salmon Growers Association (1991). Some other examples are programs of the British Trout Association, Ornamental Fish Industry (United Kingdom), Washington Fish Growers Association, Aquaculture Foundation of India, Pacifi c Coast Shellfish Growers Association, and the Association of Scottish Shellfish Growers. The Australian Prawn Farmers Association developed one of the fi rst codes of practice for shrimp culture (Donovan 1998). The Global Aquaculture Alliance (Boyd 1999) and several other organizations have since developed codes of practice for responsible aquaculture and especially for responsible shrimp farming.

Voluntary Adoption of BMPs

Because of the huge amount of attention given to the negative environmental impacts of aquaculture and promotion of better practices, there no doubt has been voluntary adoption of BMPs by individual farms. It is unlikely that voluntary adoption has been widespread or systematic, because BMP implementation depends on the individual farmer's percep-tion of the effect of adoption on farm profitability. It is usually easier to assess the costs of BMP implementation than it is to measure benefits, so farmers often have a distorted picture of the farm-level economics associated with BMP adoption. If they perceive that BMP adoption is too costly, they will not change their behavior. BMPs that are directly associated with better production efficiency (such as practices that improve feed conver-sion) will be more readily adopted than practices that are management-intensive or require investment in additional infrastructure. No studies have been conducted to determine the degree of independent, voluntary adoption of aquaculture BMPs.

RETAILER PURCHASING POLICIES

Several large retailers have developed purchasing policies and production standards for use in sourcing responsible aquaculture products. Examples are Marks & Spencer and Tesco in the United Kingdom, Carrefour in France, and Wegmans Food Markets, Inc., and Bon Appétit Management Company in the United States. Retailer programs tend to focus on the production process, product specifications, and product quality. However, they generally include environmental and social requirements. Standards may vary con-siderably among different companies. Standards may be agreed upon between buyer and producer and not disclosed to the public. Some retailers, e.g., Wegmans and Bon Appétit Management Company, apparently will disclose their policy and standards. Retailers have their own specialists to audit producers or they rely on auditing companies.

Retailer programs can be important in providing safe, high-quality products and in reducing negative environmental and social impacts. The standards, however, are devel-oped specifically for the retailers' marketing objectives and customer base. The standards may not be revealed to the public, and development of standards is not always a transpar-ent process. Some retailers are hiring professionals with experience related to aquaculture and environmental issues. Others are contracting with environmental NGOs to obtain assistance in developing purchasing policies and standards. Of course, this approach possibly also is a means of obtaining an endorsement from the NGO.

The primary purpose of retailer programs is for a company to differentiate itself by developing its own standards. This suggests that the retailer understands and controls the suppliers. The standards also suggest that the retailer is seeking a product specifically for its customers. These programs are marketing strategies intended to satisfy a company's customer base and to demonstrate that the company is concerned about the environment and social issues.

Certification

The purpose of aquaculture certifi cation is to provide products certifi ed to have been pro-duced by environmentally and socially responsible methods. There are several generally accepted requirements for developing certifi cation programs. The standards should be developed through a broad, deliberate, and transparent process that involves all stake-holders. These standards should be available to the public. Producers participating in certifi cation must comply with the standards, and compliance must be verifi ed through regular and unannounced inspections by a qualifi ed third party. Noncompliance with cer-tifi cation requirements must be corrected within a specifi ed period, or certifi ed status is revoked. Product traceability is a key feature of certifi cation because it must be possible to trace a certifi ed product back to the farm of origin.

Government Programs

The French Label Rouge program is an example of government-controlled certifi cation. This program certifi es many food products for the French market and it has standards for aquaculture certifi cation. It is run by an association of industry and certifying organizations with oversight by the French government. Like retailer programs, it focuses on product quality but it has environmental and social standards.

In Thailand, the Department of Fisheries has developed two BMP programs for shrimp culture that have been promoted (so far unsuccessfully) as a means of assuring buyers that shrimp have been produced responsibly. One is called the *Thailand Marine Shrimp Farming Code of Conduct (CoC)* and the other is known as the *Good Aquaculture Practices (GAP)* for marine shrimp farming. Both programs require implementation and verifi cation of specific BMPs but do not have clearly defi ned standards. Based on the practices required for compliance, the CoC program appears more rigorous than the GAP. The Department of Fisheries has trained inspectors for these programs and many farms have been certifi ed. Representatives of the shrimp culture industry were involved in dis-cussions related to formation of both programs, but apparently few other stakeholders had input. The method for inspecting farms for compliance with certifi cation requirements does not appear to be a true third-party system.

The CoC and GAP programs have created an interesting situation. The DOF was encouraged by the World Bank and other outside groups to develop the CoC and GAP as a proactive approach. The Thai Department of Fisheries wants its certifi cation program to be accepted by foreign shrimp buyers, and it desires sole responsibility for certifi cation of farms in Thailand. So far, foreign shrimp buyers are not willing to accept the Department of Fisheries program as a credible means of sourcing environmentally responsible shrimp. Some in the Department of Fisheries feel slighted because of the lack of confi dence in the CoC and GAP, and they have responded by not cooperating with those interested in install-ing other certifi cation programs. The FAO has announced an initiative to develop inter-national guidelines for aquaculture certifi cation. The fi rst expert consultancy was held in Bangkok, Thailand, from 27–30 March 2007, and at least three more consultancies are planned for other venues. This effort possibly will provide suggestions for inclusion of national interest in development of standards and administration of certifi cation programs.

The shrimp farming code of conduct developed in Madagascar can be considered a type of government certifi cation. Many of the practices presented in this document are legally binding (Razafi tseheno 2003; Edaly 2003). This code prohibits shrimp farms in mangrove areas, specifi es minimum distances between shrimp farms, prohibits imports of broodstock and postlarvae, places limits on production intensity, and requires effl uent monitoring. Adoption of the practices by all producers has allowed shrimp to be sold under a "Product of Madagascar" label and receive a premium price.

An Industry Certification Program

The Global Aquaculture Alliance (GAA) developed certifi cation standards for shrimp aquaculture and licensed these standards to the Aquaculture Certifi cation Council (ACC). The GAA has shrimp certifi cation standards for hatcheries, farms, and processing plants. The ACC has held several training programs to train inspectors, and many farms have been certifi ed. This program was given a great boost recently when Wal-Mart signed an agreement with ACC to sell only ACC-certifi ed shrimp as soon as the program can be widely implemented (Chamberlain 2005, 2006). The GAA currently is developing standards for certifi cation of several fish species at the request of Wal-Mart and several other buyers.

The relationship between GAA and ACC has been the subject of considerable contro-versy. The GAA was established in 1997 with the stated mission "to further environmen-tally responsible aquaculture." The GAA developed BMPs for voluntary adoption by shrimp farmers (Boyd 1999). This effort led to the idea of certifi cation. Certifi cation stan-dards were developed for which producers would have to install BMPs to achieve compli-ance. This program was called *Best Aquaculture Practices (BAP)* certifi cation. The program has been widely criticized by NGOs for not involving a wide range of stakeholders in developing the standards. This criticism is not entirely valid, for several environmentalists were invited to review the standards, but only a few did. The ACC was developed by GAA as the certifying body for the BAP program. Environmental NGOs complained that GAA and ACC were in reality the same organization because there was much overlap on the board of directors of the two organizations. The two organizations have gradually become independent bodies, and today, only one individual is a member of the board of directors of both organizations. The ACC and GAA have distanced themselves and operate independently, and the criticism, in my opinion, is no longer valid.

The ACC developed the procedures for verifi cation of compliance with BAP standards and trained inspectors for the program. The environmental NGOs correctly argue that the ACC is not a true third-party certifi cation program because the inspectors are not from an independent organization. At Wal-Mart's request, Conservation International (CI) evalu-ated the GAA standards and the ACC procedures. GAA and ACC incorporated changes suggested by CI, but the ACC certifi cation program still is viewed with suspicion by most environmental

NGOs. Originally, certifi cation was promoted as a way of obtaining a higher price for aquacul-ture products. However, Wal-Mart apparently does not plan to pay a higher price for ACC-certifi ed shrimp. The producer will have to bear the extra cost of certifi cation to assure sales to Wal-Mart. Environmentalists would like to see producers benefit from a higher price for their products. Nevertheless, they are probably willing to accept the Wal-Mart model because it will result in a much wider need for certifi cation than would result from high-end markets for certifi ed aquaculture products. Wal-Mart's competitors already are initiating efforts to market certifi ed aquaculture products. It is not yet clear as to whether they will pay producers more for a certifi ed product.

The WWF conducted a study of the issues that should be considered in developing certi-fi cation standards for shrimp, salmon, catfish, tilapia, trout, oysters, mussels, clams, scal-lops, abalone, and seaweed (Boyd et al. 2005). Reports have been prepared on the use of fishmeal in aquaculture and on the markets for certifi ed aquaculture products. Stakeholder dialogues have been conducted for salmon, channel catfish, tilapia, and shellfish certifi ca-tion issues, and additional dialogues are scheduled. The WWF received additional funding for this effort in early 2007, which has allowed them to hire several specialists to manage the dialogues for different species or species groups. The timeline for the project calls for stakeholders to reach agreement on issues and develop the initial certifi cation standards for the different aquaculture species by the end of 2009.

The WWF program emphasizes adoption of better practices to improve the performance of aquaculture (Clay 2004). Clay also suggests use of the term *better management practices* instead of *best management practices* because practices are evolving (see Chapter 2). Adoption of better practices will allow farms seeking certifi cation to comply with a system of numerically verifi able standards.

Although the WWF is devoting much effort to aquaculture certifi cation, it will not be the certifying body. A suitable organization to oversee the certifi cation program will be found and the WWF effort entrusted to it. This is the same way that the WWF and its donors operated in developing the Marine Stewardship Council (MSC) and the Forest Stewardship Council (FSC). The MSC develops standards for sustainable fisheries, but an independent certifying body is responsible for determining whether a fishery complies with the standard. This is true third-party certifi cation. The FSC operates in much the same way as the MSC. There have been discussions among the GAA, ACC, WWF, and MSC about possibili-ties for combining efforts. A certifi cation program developed and operated through col-laboration among the WWF, GAA, and ACC would probably be more widely acceptable than either an NGO or industry program alone. However, the author is of the opinion that aquaculture certifi cation and fisheries certifi cation should not be combined. Aquaculture is potentially more responsible than fishing, and the aquaculture industry should promote their products separately.

ISO Certification

Certifi cation by the International Standards Organization (ISO) has been achieved by several aquafeed plants and processing plants for aquaculture species. A few shrimp and fish farms also have obtained ISO certifi cation. There are several types of ISO certifi ca-tion; the applicable ISO programs for the aquaculture industry are 9001 (quality), 14001 (environment), and 22000 (food safety). ISO currently does not recommend aquaculture production standards, but the program provides a procedure through which producers can develop and implement practices for compliance with standards from another source. In ISO certifi cation, an organization says what it will do and then proves it does what it says it will do (von Zharen 1996). Retailers and other buyers of aquaculture products likely view ISO certifi cation favorably because it suggests that farms and processing plants are well managed. ISO certifi cation is not proof of environmental and social responsibility, because the entity undergoing certifi cation often makes its own standards without stake-holder involvement. Nevertheless, ISO certifi cation is established and respected, and most environmental NGOs probably have a positive opinion of it.

EurepGAP Certification

The EurepGAP program is promoted as responding to consumer concerns on food safety; animal welfare; environmental protection;

and worker health, safety, and welfare. This program, administrated from offi ces in Germany, was developed initially for traditional agricultural products. EurepGAP recently extended its standards to include aquaculture.

EurepGAP has an *Integrated Aquaculture Assurance* document with which all certifi ed facilities must comply. This document does not have clearly stated standards. Instead it has several general principles related to worker safety, facilities management, chemical use, animal welfare, environmental protection, product quality, and product testing. It requires the preparation of many action plans and has a huge record-keeping component. A total of 213 control points are identifi ed in the basic aquaculture assurance document, and each must be audited for compliance with stated criteria. Some control points have multiple criteria. For each individual aquaculture species or species group, there is an add-on module for specific control points. The add-on module for shrimp has an additional 32 control points. There are major, minor, and recommended control points. A facility must comply with all major criteria and 90% of minor criteria, but compliance with recommended criteria is not mandatory. Although the EurepGAP program is extremely detailed, it is much weaker on environmental and social requirements than most other programs discussed above. The requirements for written plans and record keeping are complicated. Only a large farm capable of dedicating one or more employees to the tasks could comply with the criteria. The author has not heard environmentalists comment on the EurepGAP program, but it seems unlikely that they would fi nd this program satisfac-tory for lack of social standards. Moreover, it is unlikely that producers would choose such a complex program over a simpler one unless forced to do so by buyers. EurepGAP was a presenter at the 27–30 March 2007 FAO consultancy on international certifi cation standards for aquaculture in Bangkok, Thailand. It appears that EurepGAP will develop specific aquaculture certifi cation programs for different countries. For example, a presenter from China talked about a EurepGAP-China aquaculture certifi cation program "based on the realities of Chinese aquaculture." This suggests that EurepGAP is planning to make changes in its program that will facilitate certifi cation of small-scale producers in developing countries.

Organic Certification

Organic certification of terrestrial food products has a long tradition. The primary purpose of standards is to verify that products are organically produced, but a few environmental and social standards usually are included. Organic certifi cation recently has been extended to aquaculture products by several organizations. The most popular ones are Naturland (Germany) and Soil Association (United Kingdom). The International Federation of Organic Agriculture Movements (IFOAM) developed aquaculture standards for inclusion in the IFOAM Basic Standards in 2005. The French Ministry of Agriculture and Fisheries is working on organic standards for aquaculture products, but presently organic aquaculture standards for France are based on agricultural standards. For example, OSO in Madagascar and the French auditing company Bureau Veritas developed organic shrimp standards that were validated by the French Ministry of Agriculture and Fisheries.

The United States Department of Agriculture (USDA) is developing standards for organic aquaculture under the National Organic Standard Board (NOSB). The Aquatic Animal Task Force has developed draft organic standards that have been approved by the NOSB and will now undergo interagency review. Once standards are approved, the USDA National Organic Program will publish an organic aquaculture certifi cation rule in the Federal Register. After implementation of the rule, all organic aquaculture products sold in the United States must comply with USDA organic standards.

CONSTRAINTS IN RESPONSIBLE AQUACULTURE PROGRAMS

The responsible aquaculture movement is based on the premise that consumers in develop-ing nations desire fish, shrimp, and other aquatic food products from environmentally and socially responsible sources. Therefore, retailers, restaurateurs, and other larger buyers are seeking such products. Environmental NGOs are interested in promoting environmental and social stewardship. Thus, they see aquaculture labeling programs as an effective way of promoting their policy agenda. International development agencies such as the World Bank, Asia Development Bank, and FAO are committed to environmental stewardship and programs by buyers to source responsible products that are compatible with that goal. Processors and exporters of aquaculture products would obviously participate in respon-sible aquaculture as a business opportunity. However, despite the popularity of respon-sible aquaculture, there remains perplexing issues that are seldom considered.

Extending BMP and Labeling Programs to Small-Scale Producers

Most aquaculture is conducted in developing countries, and a large amount of the produc-tion is from small farms. Although much of the production of small farms is consumed by farm families or sold in the local market, some of it enters the export market. The large demand for imports of aquaculture products in wealthy nations cannot be satisfi ed by the production of large farms alone. However, the current effort on aquaculture labeling is directed almost exclusively at large farms. The cost of implementation of practices, analyses, and record keeping necessary for compliance with certifi cation or other labeling programs and the expense of third-party inspection is more than most small-scale pro-ducers can bear.

In 2006, the GAA convened an expert consultation in Bangkok, Thailand, on issues related to certifi cation of small shrimp farms in Asia. The participants agreed that small farms need to be arranged in clusters to reduce the costs of compliance and verifi cation. Moreover, it will require much work to develop procedures for establishing and adminis-trating these clusters. The situation will differ from country to country and a single model for cluster farms would be unworkable. The participants also concurred that standards for ACC certifi cation would require modification for application to small farms. Small-scale producers would often not have title to their land, few would be able to demonstrate social stewardship, and effl uent monitoring and other analyses would be excessively expensive. Small-scale farmers may integrate aquaculture with poultry, swine, or cattle production. Water supplies for small-scale farms may be highly polluted, and farm families may reside beside ponds and discharge domestic waste into them. These scenarios raise food safety issues that must be addressed by labeling programs. The GAA has charged a committee with developing certifi cation standards for cluster farms. The objective is to require the same degree of environmental and social stewardship for small farms as expected for large ones, but this may prove an impossible task.

Some feel that environmental concerns are being favored over social issues in BMP and labeling programs. In a policy analysis of the ongoing effort to apply BMPs to shrimp farming, Béné (2005) suggests that the BMP effort is a scientifi c approach that, by basically ignoring social issues, has allowed a number of key stakeholders to refocus the debate on scientifi c solutions, thereby preventing other groups concerned with more intractable social and political issues from engaging in the policy process. Béné apparently feels that widespread use of labeling programs will further marginalize small-scale produc-ers. Although Béné does not offer any solutions, he hints that policy making in shrimp farming is too political to be reduced to the application of BMPs and other technical interventions. Béné's eloquent argument could be applied equally well to other forms of aquaculture, but it is not clear how to avoid some of his concerns and yet progress toward an environmentally responsible aquaculture. New (2003) also expresses concern that the responsible aquaculture movement will further marginalize small-scale producers. He stresses that responsible aquaculture should be profitable but also have a conscience.

Demonstration of Benefits

Programs promoting voluntary adoption of BMPs and labeling programs requiring verifi -cation of BMPs or compliance with standards are in their infancy. Studies of the adoption of these programs have not been made, but it is certain that current participation represents only a minuscule proportion of aquaculture production. Most consumers in the United States presently cannot fi nd environmentally labeled aquaculture products without shop-ping around, because few supermarkets and restaurants carry them. Moreover, there is scant documentation that adoption of BMPs and compliance with labeling programs has environmental and social benefits.

Verifi cation of environmental and social benefits would require carefully designed studies. In some places, comparisons could be made between farms participating in BMP and labeling programs and those not participating. These comparisons could include vari-ables such as efficiency of water use, fishmeal and protein use in feeds, feed conversion rate, effl uent volume and quality, wetland destruction, and social issues such as compen-sation and working conditions for workers and relationship with local communities. In other places, it may be possible to make such comparisons before and after implementation of BMP or labeling programs at a facility. However, aquaculture often is only one of several activities that can influence environmental and social conditions. In such situations, negative impacts caused by other resource users could mask positive effects that might result from better aquaculture practices.

Constraints in assessing the benefits of implementation of BMPs and compliance with labeling program standards could be overcome by use of indicators (Boyd 2005a, b; Boyd 2006; Boyd and Polioudakis 2006). Effectiveness of voluntary adoption of BMPs could be revealed by improvements in values of indicators. Indicators could reveal ranges in the use of water, land, protein, fishmeal, energy, and other resources and in the amount of wastes produced per tonne of production for different species and culture methods. Evaluation of these ranges could be used to establish minimum acceptable values of each indicator for compliance with labeling programs. Of course, social issues and some envi-ronmental concerns such as facility siting, predator control, exotic species, antibiotic and chemical use, and general sanitation possibly cannot be addressed by indicators.

Discrepancies in Programs

Differences among sites, culture systems, production methods, and species also raise issues about aquaculture labeling. With regard to sites, mangroves and other wetlands are pro-hibited in responsible aquaculture programs as sites for new farms or expansions. However, many existing farms were constructed in wetlands. Such farms obviously should be encouraged to adopt better practices. However, should they be allowed to participate in labeling programs?

The potential for water pollution varies among culture systems and increases in the following order: ponds < flow-through systems < cages and net pens. Practices are avail-able for reducing the pollution load in effl uents from ponds and flow-through systems, but nothing can be done to reduce nutrient and organic matter release from cages other than to lessen feed input. At some sites, dilution and dispersal of waste may be adequate to avoid water pollution by cage and net-pen culture. Thus, should cage and net-pen facilities be denied access to labeling programs or should those that are sited well be allowed to participate?

Water exchange has been used to fl ush ponds and improve water quality. This practice shortens hydraulic retention time and lessens effectiveness of natural processes in assim-ilating wastes. Should water exchange be strictly prohibited, restricted, allowed for emer-gencies, or allowed on a case-by-case basis in labeling programs?

The culture of some species is more likely to have serious negative environmental impacts than the production of other species. Boyd et al. (2005) suggested that contention with issues related to certifi cation of some common species or species groups would increase as follows: seaweeds < bivalve shellfish < abalone < channel catfish < tilapia < trout < shrimp < salmon.

As there are differences with respect to certifi cation issues among sites, systems, methods, and species, it is not reasonable to consider all certifi ed aquaculture products equal with regard to environmental responsibility. For example, certifi ed tilapia from ponds might have a better environmental record than certifi ed tilapia from cages. The environmental impact per unit of production might be greater for salmon cultured in cages at a certifi ed facility than for channel catfish grown in ponds at a noncertifi ed farm.

The problems of organizing small-scale farms for certifi cation and harmonizing cer-tifi cation requirements for large- and small-scale facilities has already been men-tioned. However, when compared with discrepancies related to sites, systems, methods, and species, differential treatment of small-scale producers in labeling programs would appear to be a minor issue.

The fact remains that most aquaculture in the developing world will not be included in the responsible aquaculture movement because programs are directed at facilities produc-ing for the export market. National governments should take the responsibility for regulat-ing aquaculture facilities to minimize negative environmental impacts. Moreover, social issues should also be addressed by national governments.

Unfortunately, in many nations with a large aquaculture sector, national governments have neither the resources nor the will to deal effectively with environmental and social issues. Thus, the benefits of programs for responsible aquaculture likely will be eclipsed by the negative impacts of domestic aquaculture.

Influence on United States Producers

Most foreign producers have lower production costs than producers in the United States because they have access to cheaper labor and are subject to fewer regulations. Aquaculture imports may compete directly with the products of United States aquaculture. For example, basa *(Pangasius bocourti)* and tra (*Pangasius hypophthalmus*) from Vietnam compete with channel catfish. Catfish farmers succeeded in preventing basa and tra from being sold in the United States as catfish and an antidumping tariff also was imposed. However, China has begun to ship channel catfish to the United States and governmental intervention into this trade is unlikely.

Widespread adoption of labeling programs by retailers, restaurateurs, and other buyers would require both domestic and foreign producers to comply with requirements of these programs. As a general principle, participants in labeling programs must exceed minimum legal requirements for environmental and social stewardship. Aquaculture in the United States is regulated more strictly than in developing countries. It should be much easier and cheaper for a United States producer to comply with labeling program requirements than it would for a producer in a country with few regulations. Thus, labeling should make United States aquaculture more competitive.

4

Best Management Practice Programs and Initiatives in the United States

INTRODUCTION

Best management practices (BMPs) are specific—and often detailed—protocols, practices, or procedures to manage specific operations in a socially and environmentally responsible manner, and they are typically based on risk analysis and the best available scientifi c information (Nash 2001). Usually they are a component of a larger regulatory or voluntary effort to assist farmers in achieving environmental stewardship goals of government or a producer organization. These environment-oriented BMPs may or may not contribute to the economic success of the facility.

In this chapter we summarize regulatory and other initiatives related to aquaculture BMPs in the United States. We address 1) background and progression of events for developing federal effl uent limitation guidelines (regulations) with significant supporting actions and lessons learned, 2) general description of a nationally coordinated stakeholder process that helped formulate these regulations, 3) experiences of state governments and industry initiatives with BMPs, and 4) participatory processes in regulatory development. To gather current information from the different states on the status and implementation of programs employing BMPs, we prepared and distributed an electronic questionnaire to state agencies, universities, and industry organizations during November and December 2005. Follow-up interviews and published literature filled information gaps.

BACKGROUND

The quality and safety of water sources for aquaculture operations and potential external sources of contamination are major concerns reinforcing the need to manage and protect natural aquatic resources. Management is required to maintain desired water quality within production systems and to control discharges of substances defi ned as pollutants in government regulations. The Federal Water Pollution Control Act of 1948 (33 USC §1251 *et seq.*), also known as the *Clean Water Act (CWA),* is the primary federal legislation concerning water quality. The CWA was amended significantly in 1972 (Public Law 92-500), 1977 (Public Law 95-217), and 1987 (Public Law 100-4). The fi rst amendment required a technology-based approach to discharge limits. The last amendment emphasized water quality-based effl uent limits as a basis for National Pollutant Discharge Elimination System (NPDES) permits outlined in the 1977 amendment for point source discharges into navigable waters.

The CWA prohibits the discharge of pollutants from a point source without a discharge permit. These permits are issued by the United States Environmental Protection Agency (USEPA) or a state program authorized by USEPA to administer the State Pollutant Discharge Elimination System program. Currently 45 of 50 states can issue NPDES permits after optional USEPA review. National Pollutant Discharge Elimination System general permitting regulations in 1983 defi ned concentrated aquatic animal production (CAAP) facilities subject to the NPDES permit program for controlling discharges from point sources into waters of the United States (40 CFR Sections 122 [§122.24, §122.25, and Subpart D Appendix C] and 123 [§123.25]). The NPDES permit program also applies to discharges from aquaculture projects or surface waters into aquaculture facilities to stimulate production as food inputs (Section 318 of the Clean Water Act and 40 CFR §122.25). National Pollutant Discharge Elimination System permits can be issued for an individual facility (individual permit) or numerous common facilities with similar operat-ing and environmental conditions (general permit). Permits are valid for 5 years and new requirements or modifications can be made at the time the permit is renewed for another 5-year cycle.

Water quality regulations during the 1970s and 1980s focused on controlling point source pollution through NPDES permits. Categories of CAAP facilities are defi ned as hatcheries, fish farms, or other facilities that grow or hold aquatic animals with the following criteria (40 CFR §122.24, 40 CFR Subpart D Appendix C, and 40 CFR §123.25):

1) Coldwater fish or other aquatic animals in structures that discharge at least 30 days/ year but exempts facilities that produce less than 9,090 kg/year (20,000 pounds/year) and feed less than about 2,272 kg (5,000 pounds) during the calendar months of maximum feeding
2) Warmwater aquatic animals in facilities that discharge at least 30 days/year but exempts facilities that produce less than about 45,454 kg/year (100,000 pounds/year) and facilities with closed ponds that discharge only during periods of excess runoff
3) Facilities that are determined on a case-by-case basis by the permitting authority to be significant contributors of pollution to waters of the United States.

When writing a discharge permit, the authorized state or federal authorities apply either USEPA technology-based standards for an industry-specific process, end-of-pipe dis-charge standards, or effl uent criteria; or more stringent water quality standards if the discharge is likely to affect water quality based on the designated uses of receiving waters. States are required to identify waters that do not meet water quality standards and desig-nate them as water quality limited. States are then required to establish a priority rating for such waters, and in accordance with that ranking, establish total maximum daily load (TMDL) limits for specific pollutants (McCoy 2000). States may establish stricter standards than federal requirements to maintain designated water uses and prevent water quality degradation. Water quality-based standards and requirements take precedence over technology-based effl uent limitations. The NPDES permits for aquaculture facilities are issued based on the permit writer's best professional judgment in the absence of an applicable USEPA effl uent limitation guideline for CAAP facilities.

HISTORICAL MILESTONES

The need to develop a national effl uent limitation guideline for aquaculture facilities evolved over more than 20 years. The National Research Council (NRC 1978, 1979) noted the need for national effl uent regulations for aquaculture. The passage of numerous envi-ronmental protection and marine fishery conservation laws by Congress in the late 1970s also created new regulatory challenges (Jensen, in press). Early attention by USEPA on aquaculture effl uents in 1977 and 1987 resulted in limited technical guidance to NPDES permit writers but no national regulation (Jensen 2000). The USEPA (1980) evaluated the use of aquaculture systems as a technique for municipal and industrial wastewater treat-ment. The fi rst National Aquaculture Development Plan noted the importance of federal environmental regulations and the need for applied research to reduce pollution impacts from aquaculture facilities (JSA 1983a, b). Renewed emphasis on water quality and the environment resulted in technical guidance provided to public agencies and decision makers (Oceanic Institute 1991).

Diverse stakeholders identifi ed the need to characterize effl uents, document and mini-mize environmental impacts, and improve regulations through proactive dialogue among producers, regulators and public interest groups. Without a national guideline specific to aquaculture discharges, permit requirements varied among states based on the permit writer's best professional judgment. Permits were often developed based on experiences with disparate types of industrial and municipal discharges. This regulatory system and limited scientifi c information challenged local permitting authorities and aquaculture oper-ations—especially flow-though raceway facilities and emerging net-pen operations in state waters of the Pacifi c Northwest. In response, concerned stakeholders synthesized technical progress and available data, and developed prioritized recommendations (Blake et al.1992; Jensen 1992).

To address aquaculture effl uents, a federal interagency coordinating body, the Joint Subcommittee on Aquaculture (Public Law 96-362), created a Working Group on Water Quality/Effl uents in 1991. The Working Group facilitated national coordination of coop-erative research, education and public policy activities with producers and producer associations, state natural resource agencies, relevant federal agencies, academia, and other stakeholders. The Working Group helped assess existing information and identifi ed infor-mation gaps to guide research projects in anticipation of a national discharge regulation or proactive voluntary options to strengthen environmental stewardship. The Working Group was discontinued in 1993. Coordinating leadership was then provided by fi ve United States Department of Agriculture (USDA) Regional Aquaculture Centers. The Centers support aquaculture research, development, demonstration, and extension educa-tion. They directed federally funded research to characterize effl uents, develop BMPs, and improve environmental management through better culture technologies and wastewater treatment methods (CTSA 1991, 1992; WRAC 1998; Tucker 1998; NCRAC 1997, 2001, 2006; Yeo et al. 2004). The Centers also coordinated research through an interregional aquaculture waste management initiative. Further work improved the understanding of key environmental regulatory issues, analyzed current information, and recommended options for regulatory policy with reference to design and implementation of cost-effective BMPs (Rubino and Wilson 1993).

During this period, the National Research Council (NRC 1992) conducted an assess-ment of technology and opportunities for marine aquaculture in the United States. This assessment included recommendations to develop environmentally sensitive, sustainable systems; streamline permitting process for marine aquaculture; and address the lack of a formal regulatory framework to govern leasing and development of commercial aquacul-ture in federal waters. The importance of continued research and technology development to protect national aquatic resources was emphasized in a national aquaculture research and development strategic plan (JSA 1994).

From late in the 1970s until the late 1990s, the USEPA demonstrated little interest in directing limited resources to develop a national effl uent limitation guideline for aquacul-ture because of its relatively small size and minor environmental impact compared with other point source industries. The full responsibility for issuing NPDES permits fell to the authorized states and regional EPA offi ces assigned to states lacking approved program authority.

Three events converged for USEPA to promulgate a national effl uent limitation guide-line for CAAP facilities. First, a federal court consent decree settlement obligated USEPA to develop new or revised standards for various industry categories within a specific time frame (*Natural Resources Defense Council, Inc. et al. v. Whitman*, D.D.C. 89-2980; 31 January 1992). Since 1974 USEPA had completed guidelines for more than 50 major manufacturing and chemical industry categories and was now considering smaller indus-tries. Second, a national environmental advocacy organization recommended that USEPA implement the CWA for aquaculture facilities by developing effl uent limitations (Goldburg and Triplett 1997; Goldburg et al. 2001). The third event was a public notice by USEPA, under the court consent decree settlement, that solicited comments and information for its next effl uent guidelines biennial plan. For the fi rst time, this plan included fish hatcheries and farms as one of several candidate industry categories under consideration for rule-making projects (Federal Register 1998a).

Based on public comments and limited information, USEPA decided more information was needed (Federal Register 1998b). Shortly thereafter USEPA announced a preliminary study of the aquaculture industry to evaluate current wastewater controls and the oppor-tunity for improved environmental protection (USEPA 1999a). Before completing the preliminary study, USEPA and litigants amended the court settlement agreement to develop regulatory options for certain types of aquaculture facilities (*Natural Resources Defense Council, Inc. et al.* v. *Leavitt*, Civ. No. 89-2980; January 31, 1992, as modifi ed). The USEPA launched an activity to develop pollution controls in the form of a national effl uent limitation guideline for commercial and public aquaculture operations (USEPA 2000; Kreeger 2000).

The settlement resulted in a multiyear national dialogue and formal rule-making process to evaluate options for nationally applicable technology-based performance standards for the diversity of aquaculture facilities. The USEPA fi nal rule, promulgated about 30 years after fi rst recommended in the late 1970s, established a national regulatory framework and baseline compliance criteria specifically for effl uent discharges from CAAP facilities similar to other industry categories to meet CWA requirements.

FEDERAL ENVIRONMENTAL REGULATORY APPROACH AND EXPERIENCE

The USEPA made a decision to begin rule making before completing the preliminary study. The decision to undertake formal rule making for CAAP facilities ensured that funds were available to support data collection and to complete standard and customary engi-neering, economic, environmental, and other studies. The USEPA staff had no direct experience with aquaculture. Accordingly, as is customary practice for fi nal rule develop-ment, USEPA contracted several private companies to support data-gathering and complete intensive analytical modeling.

Although USEPA could have used a traditional approach to rule development, the agency supported the assistance of the federal interagency Aquaculture Effl uents Task Force (AETF). The AETF mobilized diverse expertise nationwide to assist, USEPA in developing the effl uent regulation for aquaculture operations. The two groups shared a common goal to use science and innovation as central components of assessing environ-mental protection regulatory options. The challenge was to identify reasonable options applicable under federal rule-making procedures that recognized the continuous evolution of innovation, scientifi c knowledge, and discovery.

Interagency Coordinating Task Force Model

With the decision by USEPA to conduct a preliminary study covering all aquaculture facilities, the federal interagency Joint Subcommittee on Aquaculture (JSA), created the Aquaculture Effl uents Study Task Force (AESTF) in 1999 to assist USEPA develop industry sector profi les (Jensen 2000, 2001). The USEPA senior management fully sup-ported this innovative interagency collaboration as a means of gaining expertise about the aquaculture industry and status of pollution prevention practices. A technical consultative process was established among USEPA project managers and their contractors, AESTF leadership, and recruited individuals with specific technical knowledge. The AESTF responded to requests for information and technical reviews of draft USEPA documents. The fi rst collaborative actions were preparation of frequently asked questions and answers to educate the interested public and creation of an open-access website to post USEPA information and new developments. When USEPA decided on formal CAAP facility rule making in 2000, the AESTF was renamed the *Aquaculture Effl uents Task Force (AETF)*. The potentially broad scope of this national regulation created a rare opportunity that rallied and unifi ed diverse national, regional, and species-oriented producer associations and trade organizations. The AETF provided a coordinated national forum for dialogue among stakeholders and mobilized national expertise. It provided USEPA with scientifi c knowledge, fundamental environ-mental effects data, state regulations, technical assistance, and information about facility practices in compliance with existing state effl uent regulations. The participation of key federal agencies with regulatory authority for approving drugs (United States Food and Drug Administration) and aquatic animal health permits (Animal and Plant Health Inspection Service, National Marine Fisheries Service, and United States Fish and Wildlife Service) facilitated resolution of cross-jurisdictional issues and independent interagency legal consultations.

The AETF had as many as 11 technical subgroups composed of members with recog-nized scientifi c knowledge or extensive professional experience. Technical subgroups addressed distinct production systems and priority regulatory concerns, i.e., drugs and chemicals, economics, feeds, exotic species, and aquatic and human pathogens. The AETF voluntarily prepared scientifi c reviews or special studies on drugs and chemicals, human pathogens, and economic impact analyses (Engle et al. 2005). The USEPA and USDA also collaborated through an interagency agreement to cofund a comprehensive aquacul-ture BMP technical guidance document that is the foundation of this book.

The AETF sought collaboration to gather benchmark information and data on the current status of state regulatory permit and monitoring requirements for aquaculture facilities. This same collaboration verifi ed reporting by some state regulatory authorities that aquaculture facilities were causing impairments to water quality or designated water uses. This follow-up analysis provided a national overview on the scope and relative severity of pollution associated with aquaculture from the perspective of state regulatory agencies. In numerous cases, state reporting included only speculations about pollution from aquaculture facilities.

The credibility of AETF was strengthened by the participation of several professional societies (Aquaculture Engineering Society, United States Aquaculture Society, Fish Culture and Fish Health Sections of American Fisheries Society) that supported the process of integrating scientifi c information and technical expertise into federal rule making. Signifi cant involvement came in the form of a book published by the United States Aquaculture Society summarizing scientifi c literature related to aquaculture and the environment as a scientifi c contribution to the process (Tomasso 2002). The societies carefully avoided lobbying activities or interfering with USEPA's legal policy-making role. Establishing a clear distinction between contributing scientifi c knowledge and formulating policy created an acceptable detachment that was important to their participation.

The AETF leadership conducted a midpoint evaluation of progress toward fulfillment of its stated objectives and an assessment of overall performance and satisfaction. This internal evaluation, done among diverse stakeholders, identifi ed several areas that needed improvement, including nongovernmental organization participation, outreach projects, communication among technical subgroups, cost of meeting locations, the need for tele-conferences, turnaround time on technical reviews, farmer participation, and guidance to develop outputs. This self-evaluation exercise helped improve core AETF activities and assess overall effectiveness.

The incorporation of consensus science, accurate industry profi les, participatory stake-holder opportunities, and current state regulatory requirements helped mitigate USEPA and stakeholder polarization regarding regulatory options under consideration. The estab-lishment of professional conduct and operating standards, technical competency, and transparency of work was essential to keep USEPA actively engaged with AETF. The USEPA solicited public input and received comments from many sources during the public rule-making process. However, AETF was engaged throughout the entire rule-making process and played a significant role in contributing the collective expertise and knowledge from its technical subgroups and scientifi c reports. The USEPA acknowledged the AETF as an instrumental group that provided comments, expertise, and information (Federal Register 2004). The AETF continued communications among interested stakeholders to help develop the compliance guide for permit writers and facility operators on the CAAP facility effl uent limitation guideline (USEPA 2006).

Highlights of the USEPA Rule-Making Process

The fi rst major milestone by USEPA was publication of the proposed CAAP rule for public review and comment in September 2002 (Federal Register 2002a). To develop the pro-posed rule, USEPA used information provided by the AETF and its own research. The USEPA conducted a two-phase effort to collect additional data from aquaculture produc-ers. A screener questionnaire was used to collect general facility information from all known aquatic animal producers (Federal Register 2000, 2001; USEPA 2001a). The screener questionnaire data, AETF assistance, and the national census of aquaculture (USDA-NASS 2000) were the primary sources of information for the proposed CAAP rule. The USEPA prepared additional supporting documents and released data relating to guidance and economic and environmental impact analyses (USEPA 2002a, b, c). The proposed rule elicited 300 public comments, including form letters.

The next milestone was publication of the Notice of Data Availability (NODA) that summarized the data received and described how USEPA might use the data for the fi nal rule (Federal Register 2003a). The NODA explained an intensive second phase of techni-cal data collection through a detailed questionnaire completed by about 200 facilities (USEPA 2001b). The detailed data from select facilities allowed USEPA to revise methods to estimate costs and economic impacts. Better information helped establish baseline levels of control technologies and operational measures at aquaculture facilities. The USEPA narrowed the regulatory options being considered for the fi nal rule. The NODA also included postproposal data, results from site visits and additional sampling data. After the NODA was published, only about 20 public comments were received, signifying broader support for the direction of rule options under consideration by USEPA. Initially, there was a wide range of treatment options under consideration by USEPA, from stringent end-of-pipe numeric limits for specific pollutants to continuation of the status quo for all aquaculture production systems and facility sizes.

In developing the fi nal rule, USEPA conducted a variety of public outreach activities to solicit input and educate affected and interested stakeholders on the status of the rule-making process (Federal Register 2002b). These activities included site visits and sam-pling trips, AETF meetings, public meetings during public comment periods, participation in aquaculture conferences, meetings with other federal agencies, and posting materials on the USEPA and AETF websites.

The Small Business Regulatory Enforcement Fairness Act (Public Law 104-121) addresses the concerns and potential impacts of federal regulations on small businesses. Because the aquaculture industry is dominated by small businesses (defi ned as less than $750,000 annual gross revenue; USDA-NASS 2006), USEPA convened a small business review panel. Key participants on the panel were aquaculture producers representing diverse systems and species and representatives from several federal agencies, including USEPA. A report included comments and recommendations from aquaculture producers and the panel's fi ndings and recommendations (USEPA 2002d). Recommendations from this process and analyses influenced the facility size threshold (45,454 kg or more, annual production) established in the fi nal rule.

Federal agencies are stakeholders in federal rule making. The Offi ce of Management and Budget (OMB) oversees an internal federal interagency regulatory review and plan-ning process (Federal Register 1993). This interagency regulatory review step also engaged technical expertise and political management of different federal departments with an interest in the outcome of the fi nal rule. Numerous agencies in the Departments of Agriculture, Commerce, and Interior were actively engaged in consultations with USEPA on technical issues and OMB on policy matters; the Offi ce of Advocacy of the United States Small Business Administration also played a significant role in this process.

Finally, the CWA allows judicial review of the fi nal rule by the United States Circuit Court of Appeals. In the case of the aquaculture fi nal rule, no petition was fi led requesting court-mediated dispute resolution. The absence of a court challenge by stakeholder orga-nizations or individuals was a clear indicator of broad acceptance or indifference for the effl uent limitation guideline and concomitant regulatory oversight by authorized state governments. In cases of disputes with environmental regulations, arbitration or litigation may be an option for resolution based on the applicability of laws to the facts of a specific case. This course of action can be costly for litigants in reaching a legal outcome from extended delays through the appeals process or out-of-court settlement agreements.

The fi nal rule established a national effl uent limitation guideline and new source per-formance standards for the CAAP point-source category in commercial (for profit) and noncommercial (public) facilities (Federal Register 2004) and marked the end of a nearly 4-year rule-making process.

The USEPA Aquaculture Regulation

The aquaculture effl uent limitation guideline was published with a complete description of the applicable legal authorities, environmental requirements, and rationale for the fi nal rule (Federal Register 2004). The technology-based regulation applies to CAAP facilities with annual production of 45,454 kg (100,000 pounds) or more, with exemptions for inland pond systems, alligator farms, and open-water shellfish systems. The definition of a CAAP facility for coldwater structures with annual production of 9,090 kg (20,000 pounds) or more under NPDES permitting regulations was increased to 45,454 kg, based on economic achievability and affordability criteria for this new regulation (40 CFR §122.24 and Subpart D Appendix C). The USEPA estimated that compliance for this regulation will impact about 240 facilities nationwide. The majority of the estimated total annual compliance cost of $1.4 million falls on public hatcheries. The number of facilities in scope of the regulation is low compared to the estimated total of 4,300 aquaculture farms in the United States (USDA-NASS 2006). There are currently a limited number of recirculation-system farms above the production threshold and affected net-pen operations. Most facilities directly impacted employ flow-through raceways for production of cold-water species (primarily salmonids).

The types of production systems or subcategories were narrowed to include a combined subcategory for flow-through and recirculation systems and a separate subcategory for net pens and submerged cages. Effl uents discharged and confined to private property are exempt.

Discharges into sewers flowing into publicly owned treatment works (POTWs) are not covered because of the low mass loading of pollutants. The POTWs must develop local limits for users if the mass loading from a facility causes a violation in the NPDES permit. The effl uent limitation guideline did not recognize fish, including nonnative species, as a pollutant for regulation under the CWA and NPDES permit system.

Many point source dischargers are required to comply with numeric effl uent limits established for specific pollutants. However, USEPA recognizes that BMPs can be an important part of the NPDES permitting process to prevent releases of pollutants. Over the years, BMPs for many types of facilities (metal fi nishing; organic chemicals, plastics and synthetic fi bers; textile manufacturing; pulp and paper manufacturing; pharmaceuti-cals manufacturing; and petroleum refi ning) have been developed. Case studies have demonstrated the success and fl exibility of the BMP approach in controlling releases of pollutants to receiving waters. Pollution prevention practices have become part of the NPDES program, working in conjunction with BMPs, to reduce potential pollutant dis-charges. Pollution prevention methods reduce costs and pollution risks through source reduction and recycling/reuse techniques (USEPA 1993). The effl uent limitation guideline for concentrated animal feeding operations also included guidance for conservation and pollution prevention practices (BMPs) that may improve production efficiency and protect the nation's waters (USEPA 2004).

The core requirement for flow-through and recirculation systems is development of a facility BMP Plan that specifi es how the permittee will 1) employ effi cient feed manage-ment and feeding strategies; 2) minimize discharges of accumulated solids during routine cleaning, inventorying, grading, or harvesting aquatic animals; 3) remove and dispose aquatic animal mortalities on a regular basis; 4) ensure proper storage of drugs, pesticides, and feed to prevent spills; 5) implement procedures for proper containment, cleaning, and disposal of spilled material; 6) maintain the containment structure with inspections to identify and promptly repair damage and conduct regular maintenance to ensure proper system operation and functioning; 7) maintain records to calculate feed conversion ratios and document frequency of cleaning, inspections, maintenance, and repairs; and 8) train all relevant personnel to ensure proper clean-up and disposal of spilled material, spill prevention and emergency response, operation and cleaning of production and wastewater treatment systems, and feeding procedures and equipment use.

For the net-pen or submerged cage system category, the same annual production thresh-old of 45,454 kilograms or more applies. The exception is native species released after a growing period of no longer than 4 months to enhance commercial or sport fisheries. Facilities in this subcategory must meet the same BMPs for mortality removal, materials storage, maintenance and inspection of the production system, record keeping, and train-ing as the flow-through and recirculation system subcategory. However, feed management strategies must minimize the accumulation of uneaten food beneath pens through active feed monitoring and management practices or good husbandry practices approved by the permitting authority. Waste management and disposal specifically refer to feed bags, packaging material, waste rope, and netting. Special attention is directed to transport or harvest discharge of aquatic animal blood, viscera, and carcasses or transport water containing blood.

The aquaculture effl uent guideline applies to CAAP facilities located in the territorial seas (Mean Low Water to 12 nautical miles) and the Exclusive Economic Zone (12 to 200 nautical miles). Net-pen (or cage) facilities operating in the Exclusive Economic Zone are considered to be point sources subject to new source performance standards and NPDES permit requirements. This scope of coverage is important because there are currently no model aquaculture facilities operating within this area. The federal government has proposed a regulatory framework for aquaculture facilities in these federal waters (Bush Administration 2004; NOAA 2006).

The renewed federal interest to develop national effl uent limitation guidelines for aqua-culture gained nationwide attention. It influenced numerous states to review their CAAP facility discharge regulations and induced several producer organizations to take proactive steps to develop reasonable state regulatory options. The approach preferred by numerous states and producer organizations was a BMP program that was system-, species-, or site-specific to mitigate potential environmental risks. Some states used BMPs as a regulatory model while others used end-of-pipe numeric limitations for some production systems, such as flow-through raceways.

Numeric discharge limits for a specific pollutant of concern and BMPs are two regula-tory options that may be used to meet common CWA objectives. BMPs provide fl exibil-ity for producers to incorporate evolving science and innovation based on site-specific factors to achieve a desired environmental outcome. Numeric limits defi ne a maximum concentration or discharge of a specific priority pollutant determined by periodic monitor-ing and testing of end-of-pipe discharges. The limits can be determined based on the performance of best available technologies.

Federal Assistance for Conservation-Related Programs

Federal agencies administer a variety of fi nancial assistance programs that may assist landowners with voluntary conservation, meeting requirements of government regulatory programs and aiding government-industry partnerships to promote sound environmental management. In the case of USEPA, Section 319 of the CWA directs states to submit assessment reports that demonstrate intergovernmental coordination and public participa-tion in implementing BMPs or other management options for pollution sources. States prepare management programs for controlling pollution from nonpoint and point sources into navigable waters within the state and improving the quality of such waters. To assist states in implementing these programs, USEPA administers a grant program to provide cost-share fi nancial assistance. This can support demonstration of aquaculture BMPs or other eligible activities (training and publications, for example). Funds are available to state agencies, colleges and universities, and nonprofit organizations on programs to prevent, control, and/or abate nonpoint source water pollution. Florida has taken advantage of these funds to hold statewide producer education sessions and install educational signage that illustrates BMPs at some university aquaculture research facilities.

The USDA's Natural Resources Conservation Service (NRCS) administers the Environmental Quality Incentives Program (EQIP) as a voluntary conservation program for farmers and ranchers to promote agricultural production and environmental quality. The EQIP offers limited fi nancial and technical assistance to eligible participants to install or implement structural and/or management practices on agricultural land. Fish or other animals raised by aquaculture are included under the definition of livestock (Federal Register 2003b). The EQIP offers contracts as incentive payments and cost sharing to implement conservation practices. Conservation practices are implemented according to a farm-specific conservation plan, developed in conjunction with the producer, that addresses the landowner's natural resource goals and objectives. Landowners receiving cost-share assistance must implement practices according to NRCS technical standards and specifications adapted for local conditions. The EQIP funds are distributed through a competitive process based on state ranking criteria. These federal cost-share funds have supported limited conservation practices for shellfish farming in several New England states and installation of a wetland fi ltering system for an aquaculture facility in Wisconsin.

The NRCS has partnered with several state initiatives to develop aquaculture BMPs referenced to NRCS production codes with common interests in promoting sound conser-vation practices (NRCS 2004; LSU AgCenter 2003; FDACS 2005). The NRCS publishes conservation standards and specifications in the state Field Offi ce Technical Guide. A noteworthy spin-off by the NRCS Massachusetts offi ce was development of an interim Conservation Practice Standard, Shellfish Aquaculture Management Code 706. The purpose of the Code is to enhance the sustainability of aquaculture; minimize adverse impacts of shellfish farming on water, plant, animal and human resources; ensure depend-able quantity and quality of water to support shellfish production; and ensure adequate quantity and quality of food to support shellfish production.

Amendments to the Coastal Zone Management Act (Public Law 104-150) in 1996 allowed the Secretary of Commerce to provide development grants to states to prepare management programs. It also established a grant program for states to regulate aquacul-ture facilities. This amendment was recognized as an important national priority (NRC 1992) to support the development of policies and practices for marine aquaculture in state waters. States have the opportunity to recognize aquaculture as a priority area for inclusion in management programs with fi nancial support under provisions of this Act.

Federal programs support development of environmental codes of practice and BMPs for aquaculture in partnership with industry and university stakeholders (Howerton 2001; PCSGA 2001; WFGA 2002; Leavitt 2004; Malison and Hartleb 2005). Within the broad public interest and responsibility to regulate and enforce conservation and protection laws for natural aquatic resources, numerous government programs fund projects often through universities and producer organizations. Federal funding with state partners contributes scientifi c information and aquaculture extension programs to support development of relevant and practical BMPs as an acceptable option for policy programs and industry adoption. Public investments in science and technology, state regulatory initiatives, and continuous innovations and improvements by industry provide important baseline data, scientifi c information, and field experiences to assist governmental regulators in develop-ing aquaculture-specific effl uent regulations and BMPs.

STATE GOVERNMENT BEST MANAGEMENT PRACTICES PROGRAMS

State governments have a significant role in environmental regulatory programs, especially states authorized by USEPA to administer the NPDES permit program. In these cases facilities apply to the approved state agency for an NPDES permit to discharge into state waters. The USEPA retains an oversight role for NPDES program implementation and can interject during the issuance of a permit. Some states have extensive experience issuing NPDES permits for aquaculture facilities and acknowledge the important role of BMPs as guidance or the primary requrement for a discharge permit. In other cases aquaculture facilities may be required to have a BMP plan as one of several conditions for a discharge permit. Numerous state permitting requirements are more stringent than the national effl uent limitation guideline for CAAP facilities, especially for marine finfish aquaculture in open water net pens and enclosures.

Regional reviews on the status of state permitting for aquaculture facilities revealed considerable variations, mixed concerns, and challenges because of previous lack of a national effl uent guideline and reliance on the best professional judgment of state permit writers (Floyd et al. 1991; Davis 1993; Wypyszinski et al. 1994; Ewart et al. 1995). The NPDES permitting regulation definition of a CAAP facility in 1983 was not relevant for fl oating net pens in open marine waters. Emerging net-pen facilities were confronted with NPDES permit requirements in the late 1980s. There was no general NPDES permit template or permitting experience for these new systems in state coastal waters. This situation created significant burdens on state regulatory programs and permit applicants. The ongoing legal and regulatory environment continues to be a major constraint for marine aquaculture development (DeVoe 1999). Shellfish farming is generally considered an established, traditional activity in most coastal states and recognized as a legitimate activity in federally approved state coastal zone management plans (Nelson et al. 1999).

Five states (Arizona, Florida, Hawaii, Idaho, and Maine) had taken the initiative to use BMPs for voluntary guidance or mandatory regulations prior to 2000. The USEPA's deci-sion to develop an effl uent limitation guideline for CAAP facilities, including BMPs as a regulatory option, influenced numerous state programs and industry associations to take proactive steps using the BMP approach. Some states took a wait-and-see approach because of the uncertainty about the requirements in USEPA's fi nal rule. From the perspective of state regulatory programs, despite differences in NPDES permitting requirements, there was limited concern about significant pollution impacts from aquacul-ture facility discharges covered under existing effl uent regulations. However, the cost and complexity of permit compliance were significant to some commercial sectors and public hatcheries impacted by site-specific factors and type of production system (Floyd et al. 1991). Increased engagement by environmental advocacy organizations in some state regulatory programs and periodic cases of costly litigation (Nash 2001) created uncertainties for operational practices and fi nancial risks. Renewed federal attention catalyzed north-central and northeastern regional efforts to assess critical environmental issues and provide guidance to facility owners and regulators (Mugg et al. 2000; Westers 2003).

Twenty-one of 50 states responded to our national questionnaire soliciting current information about state agency regulatory programs and voluntary initiatives associated with BMPs and aquaculture facilities. These states represent geographical areas where aquaculture is an important activity. Numerous state agencies have developed and imple-mented programs. Other states were developing BMPs, completing general guidance, encouraging BMPs for improved production practices, or requiring animal husbandry practices to protect farms from nuisance lawsuits under an existing right-to-farm law. Several states expressed interest in developing BMPs. State regulatory programs provide models for BMPs, and in conjunction with USEPA's support for BMPs, should stimulate broader BMP application by state regulators and industry groups. Maryland passed legislation to form a coordinating council to draft BMPs for aquaculture operations. California passed legislation regulating marine finfish aquacul-ture in state waters that requires operators to establish lease-site BMPs for disease, escape, and environmental stewardship (2006 California Senate Bill 201 *Marine Finfish Aquaculture: Leases*). The following highlights programs and initiatives led by state agencies, often in partnership with industry stakeholders and academia.

Voluntary Programs

For cases in which aquaculture operations are below the thresholds for mandatory effl uent regulations, state agencies have collaborated with stakeholders to develop BMPs as guidance for voluntary adoption. These are proactive steps intended to safeguard the environment through improved environmental management practices. They also function as trials to determine effectiveness in lieu of mandatory prescriptive requirements. Many of these voluntary BMPs are relatively new, but continued monitoring of their effective-ness will likely determine long-term acceptance and broader adoption in other states.

Alabama

The Alabama Department of Environmental Management (ADEM) recommends the appli-cation of voluntary BMPs to prevent and/or minimize possible environmental impacts. Each BMP consists of a definition, explanation of why and how an environmental effect may arise, list of suggested practices or combination of practices to mitigate or prevent environmental effects, and additional information (implementation notes) to effi ciently achieve the desired environmental protection outcomes. In the event of a demonstrated in-stream water quality effect from a facility, ADEM has already determined it would use BMPs as a regulatory baseline to add new permit requirements.

Massachusetts

The Department of Agricultural Resources encourages the development of shellfish aqua-culture BMPs as an industry-driven activity. In addition to sanctioning BMPs, the depart-ment encourages broad application and implementation of these practices. The shellfish BMPs are a set of voluntary procedures focused on maintaining or increasing production while minimizing impact on the environment and improving relations with other coastal resource users. This voluntary initiative exemplifi es state-level interest (farmers and agencies) to exceed federal requirements supported by a partnership with the USDA Risk Management Agency (Leavitt 2004).

Maryland

Legislation enacted during 2005 created the Maryland Aquaculture Coordinating Council (MACC) consisting of 17 members from industry, academia, regulatory, and political organizations. Among other responsibilities, the MACC was charged with the develop-ment of voluntary Best Management Practices (BMPs) for all forms of aquaculture.

To accomplish this task, the MACC created six subcommittees chaired by MACC members, with additional membership provided by MACC members, agency and producer representatives, and aquaculture extension specialists. During the summer and fall of 2006, the subcommittees met and formulated drafts. Subcommittee meetings were open to public participation.

The BMPs consist of existing state and federal laws and regulations as well as produc-tion system designs and practices and conservation practice standards from NRCS, Southern Regional Aquaculture Center, and other aquaculture extension publications that will "min-imize, so far as practicable, pollution or environmental disruption" (MACC

2006). Topics addressed include water resources and management; production systems; shellfish culture; nonnative species; aquatic animal health; and shipping, transport, and sale. The MACC will review the BMPs and all laws and regulations pertaining to aquaculture in Maryland on a regular basis.

Michigan

The Michigan Right-to-Farm Act provides farmers with protection from nuisance lawsuits. This state statute authorizes the Michigan Commission of Agriculture to develop and adopt Generally Accepted Agricultural and Management Practices (GAAMPs) for farms and farm operations that include aquaculture species (Michigan Commission of Agriculture 2006). The GAAMPs are reissued annually with updates as needed. These voluntary practices are based on available technology and scientifi c research to promote sound envi-ronmental stewardship and help maintain a farmer's right-to-farm. Aquaculture GAAMPs are used to promote proper animal husbandry and care practices. The GAAMPs were developed with input from the Michigan Aquaculture Association, Michigan State University, and Michigan Departments of Agriculture and Natural Resources. They are revised as new scientifi c discoveries and changing economic conditions dictate. The aqua-culture GAAMPs are not related directly to the CWA or Michigan environmental regula-tions, although there were requests that water quality related GAAMPs be developed for aquaculture.

Regulatory Programs

States that employ BMPs in NPDES permits generally adhere to the CWA definition of a BMP as a schedule of activities, prohibitions or practices, maintenance procedures, and other management practices to prevent or reduce the pollution of waters of the United States. The BMPs also include treatment requirements, operating procedures, and practices to control site runoff, spillage or leaks, sludge or waste disposal, or drainage of raw mate-rial storage (40 CFR §122.2).

States with approved NPDES program authorities generally have adopted two approaches with BMPs: 1) prevent or reduce nutrients and sediments in effl uents and/or 2) address local issues that may include species-specific management practices, construction, water sources, nonnative species, genetics, aquatic animal health, mortality disposal, and chem-ical and drug use. State regulators were concerned with farmer practices that might cause degradation of public waters, soils, vistas, or unique ecotypes not easily mitigated in cases of significant pollution. Those states also created mandates for compliance monitoring and reporting. Although most of this chapter is focused on commercial aquaculture, public hatcheries have a history of developing and implementing BMPs and performance standards for hatchery operations to comply with discharge regulations and minimize discharges (USFWS 2000).

Idaho

Idaho is one of the fi ve states not approved by USEPA to administer the NPDES permit program. The USEPA Region 10 has developed a general permit for aquaculture facilities in Idaho to satisfy the NPDES permit requirement (USEPA 1999b).

The Idaho Department of Environmental Quality (IDEQ) performs much of the on-the-ground work under con-tract to USEPA that includes facility inspections. A central feature of the general permit is a requirement for producers to develop a waste management plan with BMPs to prevent or minimize the generation and discharge of wastes and pollutants and ensure disposal or land application of wastes in such a way as to minimize adverse environmental impact. Pollution prevention and recycling are encouraged. The USEPA specifically advises permit applicants to create and adopt BMPs to address capture and disposal of solids and bio-logical wastes; detoxifi cation of chlorine or other chemicals; feed management; drug, disinfectant or other chemical storage, use or disposal; pond or raceway cleaning; main-tenance to prevent overflow or bypass events; and staff training.

University of Idaho Extension, Idaho Department of Agriculture, IDEQ, and the aqua-culture industry developed a publication to assist aquaculture (trout) producers develop waste management plans (IDEQ, undated). The guidelines provide design criteria for construction of waste management systems; owner, operator, agency and community education relative to aquaculture waste management systems; and BMPs to meet the NPDES general permit requirements. The USEPA advises producers that their BMP plan must be consistent with the general permit and guidelines.

Maine

In 1994, the University of Maine suggested biosecurity audits might be helpful to the farmed salmon industry to reduce disease risk. This introduced the concept of collective best management practices (BMPs) to the industry. In cooperation with the university, an auditing program was initiated in 1995. In 1997 the Maine Aquaculture Association (MAA) developed a code of containment that established standards for equipment and operations designed to reduce the escape of farmed fish. That code ultimately formed the basis of the Containment Management System (CMS) (Ostergard et al. 2002). The CMS is an externally audited set of BMPs embedded in a hazard analysis and critical control point-based risk management system designed to minimize fish farm escapes. The lease contracts of the Maine Departments of Environmental Protection (DEP) permit and Marine Resources require growers to submit a CMS plan for agency review and approval. The DEP permit also requires an annual audit of each farm for CMS compliance.

In 2001, the Maine Department of Environmental Protection was authorized by USEPA to administer the NPDES permitting program. During 2003, the Board of Environmental Protection promulgated a salmon aquaculture general permit containing extensive moni-toring requirements and water quality standards. The general permit includes a husbandry practice BMP and specific BMPs for facility operation, disease control, and spill control and prescribes BMPs in a CMS . The husbandry BMP specifi es stocking densities and site fallowing. Facility operation BMPs include mortalities, chemical and solid waste, biofouling, predator nets, mooring systems, and unusual events that may lead to significant environmental impacts. The disease control BMP includes drug labels, drug use notifi ca-tion, investigational new animal drugs, drug discharge, and facility signage. Spill control BMPs are prescribed within a spill prevention and control plan that includes petroleum and hazardous chemical inventory, discharge and cleanup procedures, equipment inven-tory, spill event reporting, employee training, and federal regulations.

Florida

In 1999, a state statute and associated administrative rule established an application pro-cedure for an annual aquaculture certifi cate of registration and BMPs to establish minimum standards to protect and maintain water quality, fish, wildlife, and natural systems. Producers must apply for a certifi cate from the Florida Department of Agriculture and Consumer Services (FDACS) and agree to abide by the BMPs or acquire all other appli-cable permits from government agencies. A manual of BMPs was developed through advisory committees and a public rule-making process. The manual is a dynamic document with four editions implemented over 7 years to accommodate technical and scientifi c advancement. The manual includes BMPs for water resources, construction, nonnative and restrictive nonnative species, marine shrimp culture, sturgeon culture, live rock culture, aquatic plants, health management, mortality removal, preventing wildlife depredation, shipment, transport and sale, aquaculture chemical and drug handling, and aquatic animal welfare (FDACS 2005).

The BMPs in the manual are used to preserve environmental integrity while eliminating cumbersome, duplicative, and often confusing environmental permitting and licensing requirements. The practices were developed specifically for Florida aquaculture and rep-resent a mutually beneficial relationship between commercial aquaculture production and natural resource protection. Florida aquaculturists following the BMPs meet the minimum standards necessary for protecting and maintaining offsite water quality and wildlife habitat. Farms certifi ed through this program are presumed to be in compliance with state groundwater and surface water standards. The Florida Department of Environmental Protection is not authorized to institute proceedings against any certifi ed facility to recover costs or damages associated with groundwater or surface water contamination or the evaluation, assessment, or remediation of contaminated groundwater or surface water. The FDACS enforces and verifi es BMP implementation through unannounced farm inspec-tions. Farms that do not implement or maintain the BMPs are violating state law and are subject to written notice, fi nes, suspension/revocation of the aquaculture certifi cate of registration, and potential misdemeanor charges.

New Jersey

In 1993, the New Jersey Department of Agriculture (NJDA) and producers began the process of developing agricultural management practices (AMPs) for aquaculture to anticipate environmental regulations and take proactive steps to protect the environment and conserve wild stocks (Rutgers University 2004). The NJDA is the lead agency for AMP development, approval and oversight. The AMPs are goal-oriented management practices linked to acquiring and retaining a compulsory aquatic farmer license. The prac-tices assist farmers in complying with environmental regulations and facilitate the permit-ting process. They emphasize good production practices that can yield economic benefits and improve profits. The AMPs can protect the natural environment and native fish stocks, reduce disease and mortality, improve the quality of farm-raised products, potentially lower the cost of permitting, and reduce the cost of remediation and rehabilitation. Many of the core concepts to develop AMPs were based on New Jersey Department of Environmental Protection permit requirements. No aquaculture facilities in New Jersey meet the facility requirements for an NPDES permit under the national effl uent limitation guideline.

Texas

During 1998, the Texas Commission on Environmental Quality (TCEQ) approved a general permit for aquaculture discharges into or adjacent to water in the state by all aquatic animal production facilities. Facilities that meet the definition of a CAAP under the NPDES permit regulation and coastal shrimp research facilities that exceed specifi ed discharge limits must obtain an individual permit. The general permit includes BMPs to abate the discharge of suspended solids or other pollutants (TCEQ 2006). The BMPs were developed as an outcome of administrative hearings that included input from producers, Texas Parks and Wildlife, and Environmental Defense. The BMPs include treatment requirements, operating procedures, and practices to control site runoff, spillage or leaks, sludge or waste disposal, drainage from raw material storage, or the abatement of nuisance odors and conditions. The BMPs are reasonable and necessary measures to achieve per-formance or quality standards. Furthermore, they are intended to protect air quality, surface water quality, and potential uses of groundwaters. Through implementation of the BMP plan, the permittee is expected to prevent or minimize the generation and release of pollutants to waters of the state.

The TCEQ also identifi es management practices and standard operating procedures that should be addressed: 1) modification of equipment, facilities, technology, and procedures; 2) improvement in management or general operation of the facility; 3) inspections and records; and 4) reporting of BMP incidents. Regional offi ce guidance includes specific BMPs to control solids; storage of drugs, pesticides, and feed; spill management; structural maintenance; record keeping; training; and pond aeration. While there is no formalized public process to create or modify BMPs, modification of the BMP plan by the applicant is required with any change in the facility or facility operation or when TCEQ fi nds a failure to achieve the general objective of preventing or minimizing the generation of pollutants and their discharge. The TCEQ's regional offi ces complete annual inspections to examine BMP plans and verify implementation and compliance. The Texas A&M University Sea Grant College Program published an in-depth regulatory guide to clarify the complex regulatory environment (Treece 2005).

Arizona

The 1986 Arizona Environmental Quality Act was enacted to protect surface and ground-water quality from point and nonpoint source discharges following discharge permit administration approval by USEPA. Arizona has focused on decreasing nitrate concentra-tions in groundwater as the primary objective of its groundwater pollution prevention programs. In state legislation, aquaculture is recognized as a component of agricultural production and a potential source of nitrate contamination to groundwater.

Aquaculture operations are defi ned as concentrated animal feeding operations, which are regulated to minimize environmental effects. Arizona provides industry-wide regulation through the use of agricultural general permits. These permits allow farms to operate and discharge, without an individual NPDES permit, when the operation implements BMPs. These prac-tices may be applied before, during, and after discharges to reduce or eliminate the intro-duction of pollutants into receiving waters. Economic, institutional, and technical factors are considered in developing BMPs.

Arizona's BMPs consist of three general goals that address the importance of properly 1) harvesting, stockpiling, and disposing of animal manure; 2) controlling and disposing of nitrogen contaminated water; and 3) closing operations. A variety of guidance practices (GPs) provide owner/operators with methods to implement the BMP goals. The GPs rep-resent state-of-the-art technologies available to the owner/ operator. The actual GPs chosen for adoption may vary because their environmental effectiveness depends on such factors as soil type, available farm equipment,. water quality impacts, land ownership, facility siting, and related economic criteria. The aquaculture GPs were developed by the University of Arizona and then reviewed by representatives from the Arizona Department of Environmental Quality, producers, and Arizona Department of Game and Fish (Fitzsimmons 1999).

Hawaii

Hawaii is an approved state by the USEPA to enforce the CWA. The authorized agency for granting the permit in Hawaii is the Department of Health. Following a 6-year effort funded by the USDA Center for Tropical and Subtropical Aquaculture to characterize effl u-ent discharges, the Center developed a BMP manual (Howerton 2001). The BMPs are simultaneously voluntary and a compulsory part of two permit processes. The BMPs are voluntary in that the state extension/advisory program recommends their use to farmers for farm operations. The BMPs are required in the issuance of the NPDES permit by the Department of Health and the issuance of the conservation district use permit by the Hawaii Department of Land and Natural Resources (DLNR). In the case of the DLNR, they are called Management Plans. With respect to the mandatory BMPs, the Departments of Health or Land and Natural Resources use previous experience and work interactively with permit applicants to develop the permit BMPs on a project-by-project basis. The process of nego-tiating required BMPs with permit applicants has become more effi cient. The BMPs are more relevant, reasonable, and realistic; agency personnel gained experience with aquacul-ture and farmers gained experience with what agency personnel wanted. Required BMPs are enforced through the permit and ongoing project monitoring and reporting to the regu-latory agencies. Site inspections are carried out to review farm operations.

Missouri

During 2003, the Missouri Department of Natural Resources added a requirement that farmers develop and implement BMPs for general NPDES permits required for coldwater and warmwater species. One permit covers impoundments, tanks, or recirculation produc-tion systems, and another addresses flow-through systems. Initially, the Missouri Aquaculture Council introduced the concept of adding BMPs to the general permits. The council is composed of state agencies, industry, and university representatives and cochaired by the Missouri Departments of Agriculture and Conservation. After some discussion, the permits were drafted .and made available for public comment.

The BMP requirements for both permits instruct the farmer to develop an operation and maintenance plan that employs BMPs for preventing the discharge of solids from the facility. The plan includes a list of management practices on reducing solids discharge, drug and chemical usage, and feed and fertilizer storage (MDNR 2002a, b). The Department of Natural Resources has the authority to inspect farms for permit compliance.

Oklahoma

Aquaculture facilities defi ned as CAAPs under the NPDES permitting regulation must acquire a general permit from the Oklahoma Department of Environmental Quality and use BMPs. Twenty-three aquaculture BMPs were developed based on university research and demonstration projects (Langston University 2002). The BMPs are intended to minimize pollutant loading to aquatic environments and include treatment requirements, operating procedures, and practices to control plant site runoff, spillage or leaks, sludge or waste disposal, or drainage from raw material storage. The BMPs are focused on minimizing nutrient and sediment discharges by specifying practices to reduce discharge turbidity; improve sediment settling; keep discharge and drug use records; create settling basins; reuse collected sediments;

prevent erosion (pipe placement and revegetation); use in-pond harvest basins; manage and store stormwater, fertilizers, and feeds; aerate; and harvest. Operators of aquaculture facilities who do not require NPDES permits or monitoring must adopt seven prescribed BMPs and may choose any combina-tion of the remaining BMPs, or devise their own, to achieve the goal of environmental protection.

PRODUCER INITIATIVES

State and regional producer organizations have developed codes of practice and BMPs as proactive steps for responsible environmental stewardship. These actions demonstrate commitments to environmental stewardship for purposes of minimizing potential adverse impacts from aquaculture production. Producer initiatives gained credibility through relevant research and strong backing by aquaculture producers. Signifi cant farmer input has been critical to create BMPs that are practical, effective, and minimize unreasonable costs. Aquaculture associations in Alabama, Alaska, Arkansas, Maine, and two regional shellfish farmer organizations sought partnerships with Land Grant universities or Sea Grant college programs to collaboratively develop BMPs.

Quality Assurance Programs

Development of voluntary quality assurance programs was an early initiative of several producer associations. Catfish, trout, and hybrid striped bass associations developed quality assurance programs with BMPs as guidance to prevent potential hazards related to food safety (Brunson 1993; USTFA 1994; Jahncke et al. 1996). A generic shellfish quality assurance program addressed BMPs in the context of shellfish issues (Canzonier 2002). Producer associations administer these programs as a benefit and service to members. These food safety-oriented initiatives became less important after the United States Food and Drug Administration required seafood processors and importers to comply with man-datory seafood Hazard Analysis Critical Control Point (HACCP) regulations (Federal Register 1995). However, these programs trained producers to adopt good practices in the chain-of-custody from farm to processor as the fi rst critical control point for unprocessed farm products.

Alabama Catfish Producers

Alabama Catfish Producers voluntarily partnered with the Alabama Department of Environmental Management (ADEM) and NRCS to formalize voluntary BMPs that were in use and to develop new BMPs that addressed issues identifi ed by Auburn University during an environmental impact assessment funded by the Alabama Catfish Producers.

Auburn University researched, developed, evaluated, or updated BMPs that would avoid or minimize adverse environmental impacts. Producers, Auburn University research and extension personnel, ADEM, USEPA, NRCS, Alabama Farmers Federation, and various other individuals reviewed the initial practices. The draft BMPs were formally presented to the Alabama catfish farmers to gather additional comments. After revision, the fi nal BMPs were produced as 21 NRCS conservation practice guide sheets (Boyd and Hulcher 2001; Boyd et al. 2004).

Alaskan Shellfish Growers Association

During 2001, the University of Alaska and the Alaskan Shellfish Growers Association drafted voluntary BMPs to ensure that operations were conducted to protect the natural marine environment to the greatest extent possible. This includes minimizing adverse environmental effects and maximizing beneficial effects, and minimizing impacts of shellfish farming operations on neighbors and other users of Alaska's marine waters. The environmental code presents objectives, strategies, and performance measures to address and mitigate potential adverse environmental impacts from shellfish aquaculture and pro-vides the means for monitoring compliance in implementing these strategies.

Farmers are encouraged to "take stewardship into their own hands and go above and beyond regulatory requirements" by improving upon and adding their own policies to enhance the value of the code to their operations. The code consists of fi ve sections:

> introduction; environmental health; general shellfish aquaculture practices; integrated pest management and predator control practices; and shellfish cultivation operations and prac-tices. Each section provides in-depth discussion for appropriate BMPs and a rationale for

their implementation. The BMPs are voluntary and not required by Alaska regulatory agencies.

Arkansas Bait and Ornamental Fish Growers Association

In 2001, the Arkansas Bait and Ornamental Fish Growers Association, in cooperation with the University of Arkansas at Pine Bluff (UAPB), developed BMPs for bait and ornamen-tal fish farms (ABOFGA, undated). The impetus for producing voluntary BMPs was a concern by producers that additional regulatory and monitoring costs would further con-tribute to declining numbers of fish farms in the state, and growing member awareness that larger environmental impacts are possible from small local sources.

The BMPs focus on water conservation, erosion prevention, and solids and nutrient management. In-depth narratives for each issue describe current, sensible practices and suggestions to conserve water and prevent off-farm releases of solids and nutrients. The BMPs are summarized, and a signature block includes the statement that the farmer agrees to voluntarily implement the practices.

Technical assistance is available from UAPB extension specialists to implement BMPs but there is no formal process to verify implementation. Financial assistance is not available for implementation. Anecdotal evidence suggests a high implementation rate, especially for practices with economic benefits (water conservation). Extension specialists have field-tested and proved that some BMPs are not appropriate and no longer recom-mend their implementation or have modifi ed recommendations for specific conditions. As examples, the BMP to plant grass along pond edges to reduce erosion is no longer recom-mended because this practice created habitat that favored aquatic snails, which serve as intermediate hosts for several fish parasites (N. Stone, University of Arkansas-Pine Bluff, personal communication, December 2005). Specifi c site analysis is also needed before recommending the use of vegetated ditches to capture suspended sediments (Frimpong et al. 2004).

Maine Aquaculture Association

The Maine Aquaculture Association (MAA) developed a code of containment that established BMPs and standards for equipment and operations designed to reduce the escape of farmed fish. That code ultimately formed the basis for the Containment Management System (CMS). In 2000 MAA hosted an international meeting of growers' associations focused on codes of practice, BMPs, and their design and implementation. This meeting and review of codes of practice formed the basis for the MAA code of practice (MAA, undated). This is an overarching code for compliance by all association members. It is linked to two species group subdocuments, the MAA finfish bay manage-ment agreement (MAA 2002) and the MAA shellfish biosecurity and shellfish health agreement. The latter document is still under development.

The MAA's original BMPs, the CMS, and industry code of conduct were developed in cooperation with state and federal agencies and environmental nongovernmental organizations. The MAA Code of Conduct was developed by MAA staff and reviewed by members and external reviewers including the United Nations Food and Agricultural Organization and World Wildlife Fund (MAA, undated).

The MAA writes BMPs and works to get members vested in their implementation by 1) involving members in BMP technical drafting, 2) conducting member BMP training seminars and educational meetings, and 3) working with regulators to ensure that BMPs in regulations actually achieve intended effects. The MAA also defi ned key terms prior to developing codes, systems, and practices. The BMP review and revision are a continual process driven by producers. The MAA has a finfish bay management committee and a shellfish working group to review, revise, and recommend BMPs. A steering committee advises on the implementation of the CMS. This steering committee consists of all relevant state and federal agencies, the industry, Conservation Law Foundation, Trout Unlimited, and the Atlantic Salmon Federation.

Washington Fish Growers Association

The Washington Fish Growers Association (WFGA) developed BMPs to foster and main-tain high facility operating standards in the salmon net-pen industry (WSGA 1991) fol-lowed by a voluntary code of conduct (WFGA 2002). The BMPs stressed best practices related to introduction of fish stocks, feed and feeding, mortality removal and disposal, net cleaning and changing, predator control, harvesting, medical and chemical application, spill prevention and chemical storage, solids waste collection and disposal, sanitary facil-ities and waste disposal, maintenance and repair of facility and equipment, environmental monitoring and reporting, navigation aids, safety, aesthetics and noise, and employee training and public education. The codes of practice were developed under the code and guidelines for responsible aquaculture and practices of the Food and Agriculture Organization of the United Nations (FAO 1995, 1997). The code was developed and published in collaboration with and fi nancial support from the Department of Commerce, National Oceanographic and Atmospheric Administration (NOAA).

Pacifi c Coast Shellfish Growers Association

The Pacifi c Coast Shellfish Growers Association (PCSGA) established an environmental management system for West Coast farmers. Growers recognized that increased scrutiny of their operations, due largely to the Endangered Species Act and Essential Fish Habitat rules under the Magnuson-Stevens Fishery Conservation and Management Act (Public Law 65), had the potential to impact their farming operations. Consumers are also increas-ingly demanding environmentally sustainable seafood. The PCSGA's environmental management system was designed to address these concerns (Dewey 2000).

The PCSGA's approach included an overarching environmental policy (general operat-ing principles) and detailed environmental codes of practice with sets of recommended BMPs for culture activities associated with molluscan species produced on the West Coast. The process for developing the policy and codes involved grass-roots level engagement. Eight shellfish growing areas in Alaska, Washington, Oregon, and California were desig-nated as regions for the purpose of holding small group growers meetings to develop the environmental policy followed by environmental codes and BMPs.

The policy consists of fi ve components: 1) environmental stewardship and responsible management, 2) environmental excellence, 3) regulatory compliance, 4) waste manage-ment, and 5) sharing resources (PCSGA 2001). With the environmental policy as a foun-dation, the environmental codes and BMPs were developed with the goal of minimizing adverse environmental impacts and maximizing benefits to the extent possible, utilizing the best available science. The codes were developed by a steering committee with one representative shellfish farmer from each of the eight regions. The steering committee and the regions also included Sea Grant College Program representatives. During a 2-year effort, culture practices and their potential effects on the environment, positive and nega-tive, were identifi ed. A list of recommended BMPs was developed and designed to mini-mize negative impacts and magnify positive impacts. Experts in marine biology, regulation, and policy from the Pacifi c Shellfish Institute, PCSGA, and state Sea Grant Programs drafted documents. These drafts were reviewed by each region and revised to accurately represent regional shellfish production practices and environmental conditions. The initial BMP draft was reviewed by federal and state resource agencies and environmental nongovernmental organizations.

The BMPs adopted by the PCSGA consist of 13 sections, including the purpose of environmental code of practices; background on the shellfish industry; shellfish aquacul-ture interactions in the marine environment; shellfish aquaculture operations; general management principles; pest, predator, and disease control; integrated pest management; hatchery and nursery operations; cultivation operations and practices; processing and ship-ping operations; industry regulations; a template for individual farm management plans; and a research bibliography. The environmental code of practice includes a template farm plan designed to simplify growers' development of their own farm plans. Currently, the BMPs are voluntary.

East Coast Shellfish Growers Association

The East Coast Shellfish Growers Association (ECSGA) supports the need to adopt vol-untary BMPs that minimize negative interactions with the environment and other user groups and prepare growers to deal with natural disasters. The BMPs outline the benefits and impacts of shellfish cultivation and where these methods are most appropriate. The BMPs help growers and managers address animal husbandry issues such as disease and predator control. The ECSGA is collaborating with the USDA Northeastern Regional Aquaculture Center on a project to develop an environmental code of practice and BMPs for shellfish growers on the East Coast that will also engage extension specialists in the region similar to work led by PCSGA for shellfish growers on the West Coast.

UNIVERSITY INITIATIVES

The Louisiana State University Agricultural Center (LSU AgCenter) initiated discussions with the NRCS and Louisiana Farm Bureau to develop BMPs prior to the imposition of total maximum daily loads as required by the CWA and other regulations. Several agri-cultural commodities were identifi ed for BMP development, including aquaculture. The LSU AgCenter led a collaborative effort with input from NRCS, Louisiana Departmentof Environmental Quality, Louisiana Farm Bureau Federation, and the Louisiana Department of Agriculture and Forestry to produce a BMP manual (LSU AgCenter 2003). The manual includes eight chapters: introduction, finfish pond production, crawfish pond production, crawfish nutrient management, intensive production systems, soil and water management, pesticide management, and general farm BMPs. The soil and water manage-ment chapter includes references to specific NRCS production codes for each BMP. Information about this program was distributed to the aquaculture farming community through industry newsletters, producer meetings, and website updates.

The BMPs are voluntary and can be verifi ed through farm visits by LSU extension specialists. The LSU AgCenter is under contract with the Louisiana Department of Environmental Quality to develop in-depth written guidelines, fact sheets and how-to publications to assist crawfish farmers with the adoption of BMPs. The BMPs will be revised and updated by the LSU AgCenter based on recommendations by the original participants and an initial recommendation to revisit the BMPs after 5 years. The LSU AgCenter also conducted research to support some of the BMPs, especially for crawfish aquaculture (Parr 2002).

The National Sea Grant College Program provided funds to develop an aquaculture BMP educational manual. The manual offers guidance for current and prospective aqua-culturists and regulatory agencies in Wisconsin and the Great Lakes region. The BMPs are defined as management guidelines or approaches designed to minimize or prevent any adverse environmental impacts, maximize the health and well-being of the organism being raised, and encourage effi cient and economical production. A small team of authors wrote an in-depth narrative with BMPs highlighted throughout the text. The manual consists of nine chapters: overview of aquaculture; water; management and beneficial reuse of aqua-culture wastes and effl uents; fish health; aquaculture and fish biology, species, strains and genetics; interactions with nonfish species; flow-through systems; recirculating systems; and pond systems. An appendix includes current Wisconsin environmental, resource, and fish health regulations. A committee composed of industry, regulatory agency, academics, and environmental organization representatives, provided review and comments. Aquaculture specialists are advised to add electronic copies of their respective state regulations for distribution across the region (Malison and Hartleb 2005).

CREATING A PARTICIPATORY PROCESS OF REGULATORY DEVELOPMENT

An underlying goal of participatory regulatory development must be to actively recruit and mobilize diverse stakeholders who will identify problems, participate in the regulatory development process in good faith, and offer solutions for the broader public benefit. A constructive, holistic, multisectoral dialogue about an environmental issue can improve understanding of uncertainties from different perspectives and clarify options for decision making (Luoma and Löfstedt 2007). However, when environmental issues are complex, unconstructive advocacy, a narrow focus, and exclusion of selected parties from decision making can erode public trust in science and lead to cynicism about the policies of government and the private sector (Luoma and Löfstedt 2007).

Full consensus on highly charged and emotional issues is difficult when widely oppos-ing views are held or divergent solutions are proposed. Disagreement is commonplace in public policy forums with complex environmental dilemmas, but trust may be best served by a focus on why disagreements exist rather than who is right (Luoma and Löfstedt 2007). Input of diverse opinions is an essential part of participatory policy making if the process is to be effective and legitimate. Divergent thinking assures consideration of a range of solutions and can provide opportunities for creativity and innovation (Joldersma 1997). Involvement of interest groups with divergent points of view also legitimizes the process by assuring the public that a broad range of stakeholder positions were considered during deliberations.

This is especially true for environmental issues like the USEPA effl uent guidelines effort that involve complex technological or socioeconomic questions. The successful development and broad acceptance among stakeholders of environmental codes of prac-tice, regulatory BMPs, or voluntary guidance for environmental stewardship relied on transparent stakeholder participation. Creating stakeholder respect for a transparent deci-sion making process facilitates compromises that can mollify divisive ideologies. Those affected by policies or regulations are more likely to work with regulators when they are comfortable with the working relationship and perceive that everyone is working toward a mutual and well-defi ned goal with clear criteria.

Stakeholders in Regulatory Development for Aquaculture

Stakeholders in the USEPA rule-making process used to develop an environmental regulation of aquaculture effl uents included affected producers, various federal and state regulatory agencies, scientists, and environmental nongovernmental organizations (NGOs). In the sections below, the perspective and role of each stakeholder is described, with emphasis on identifi cation of areas of divergent interests or disagreement among stake-holders. In the AETF approach—as in nearly all participatory rule-making processes— each stakeholder operated in its own self-interest, but remained open to consider perspectives of other participants in the process. This rationality, which is called "induce-ment of shared perspective" by Joldersma (1997), is essential for convergence on true consensus.

No single stakeholder can adequately provide the resources, technical expertise, or regulatory authority needed to develop a balanced and effective outcome. Partnerships of stakeholders facilitate the identifi cation of key strategic actions and the means to leverage limited resources from different programs for the broader public good.

During nearly every effort to develop aquaculture BMPs at the state level, affected producers, regulators, uni-versity scientists, environmental organizations, and the public created partnerships to gain the needed expertise and engender broad support.

Recruiting experts and other stakeholders to openly share their knowledge and field experiences can be challenging for national regulatory initiatives. Individuals may have working, professional, or statutory responsibilities to public or private programs or a paying clientele. Without direct monetary compensation, the incentive for some stakehold-ers to participate voluntarily may be driven by 1) relative importance of the issue (locally and nationally); 2) sense of peer support; 3) status of participating organizations; 4) prior commitment of in-kind resources; 5) perceived or actual risk of liability; or 6) confi dence in the organizational structure, process, and leadership. During the development of national effl uent guidelines, dozens of scientists and stakeholders contributed considerable time and effort to attend meetings, develop industry reviews, respond to USEPA questions, or otherwise participate in the rule-making process.

Industry representation

Most industry organizations consider participating in policy decisions as an interest group as an essential, if not primary, function of the organization (Furlong and Kerwin 2005). The incentive to organize is based on the enhanced political power of a large group relative to the diffuse influence of individuals. When governmental entities consider developing new or revising existing environmental regulations or policies, the stakeholder community most often directly impacted includes producers and companies whose practices or technologies are under scrutiny. Environmental regulations are often perceived by this stakeholder group as an unnecessary fi nancial burden. In the case of pollution-control regulations, producers are in essence asked to internalize the cost of pollution control in situations where that cost was previously borne by society at large. Thus, producers are inclined to resist change because regulation is perceived to adversely affect profitability, particularly in business situations were profit margins are small. Most United States aquaculture is conducted in earthen ponds that have historically been viewed by environmentalists and aquaculture scientists as having little negative impact on the environment, especially relative to other land uses and industries. Producers using pond systems perceived that criticism of the environmental performance of aquaculture was unfair because their record of environmental stewardship was good. Other producers believed that existing regulations, particularly those in certain states, were already adequate to address concerns about aquaculture effl uents.

Despite these reservations, producers, as represented by leaders of producer associa-tions, were actively engaged in the AETF-sponsored process. Producer organizations representing aquaculture are smaller and much less well developed as political interest groups than organizations representing terrestrial agriculture. With few exceptions, these organizations consist mainly of farmers growing a particular species or species group.

The various groups representing aquaculture producers rarely interact. The initial broad scope of the USEPA national effl uent guideline created a rare event that unifi ed the diverse United States aquaculture industry because of a common issue. More than ten regional and national aquaculture associations had representation throughout the entire USEPA regulatory process. They played critical roles in communicating progress and needed actions to their respective members as an important outreach function. All appointed at least one member to serve a liaison role with the AETF.

In numerous cases, there are strong local (state), regional, or national producer asso-ciations that have fi nancial resources and the will to be actively engaged in public policy processes. However, this is not always the case. Critical representation may be diminished by logistical and fi nancial constraints in prolonged, costly policy-making processes and distant meeting locations. National issues may receive less attention compared to local issues because of a poor understanding of potential direct on-farm impacts and uncertainty of how to effectively participate in a national public policy process. Local issues are best addressed by local producers and associations, with support from local programs and backing from regional or national producer associations as needed. In numerous states, producer associations play key roles in developing BMP regulatory and voluntary programs with their experience and knowledge as key ingredients for success. Innovation is a continuous process of on-farm experimentation and pioneering discov-eries that are critical for today's producers to operate successfully within a mix of eco-nomic challenges, including compliance with environmental regulations. Most small- and medium-size farm operations are fully occupied by day-to-day issues and have limited resources of time and fi nances to expend on participation in a regulatory development process. However, proper representation of farm-level practices, conditions, and experi-ences are critical to accurately understand the on-farm consequences of options under consideration by regulatory authorities and other interested stakeholders. Producers are needed to articulate the reasonableness and economic achievability of compliance alterna-tives and propose recommendations based on their understanding of a regulatory objective and knowledge of farming practices and technologies. Ultimately, producers are the envi-ronmental stewards who rely directly on the well-being of aquatic resources for their livelihood. They have daily managerial roles over the use of best practices and taking responsible actions to help sustain the environmental integrity of aquatic resources for the broader public good.

Federal agencies

The role of federal agencies in a multistakeholder process of regulatory development is central and straightforward. Agencies of the United States government are responsible for administering programs and enforcing regulations within their domain. With respect to the desirability of regulations, federal agencies are subject to policy direction from elected and appointed offi cials of the executive branch of government. Depending on the policy inclinations of these offi cials, the federal government has imposed a greater or lesser regulatory burden on affected businesses and industries as a whole. Ideally, the role of federal agencies is to balance the concerns of all stakeholders by developing regulations within existing policy frameworks that, in the case of aquaculture effl uents, provide effec-tive environmental protection without an unreasonable burden on small businesses. The role of USDA and NOAA representatives was critical to the success of the AETF process, primarily acting in key leadership roles as facilitator and catalyst. In the AETF process, the federal government, specifically the USEPA, was acting in response to a need to satisfy the terms of the Consent Decree with the NRDC to enforce provisions of the Clean Water Act, so the threat of further legal action was a strong incentive to develop a meaningful regulatory process that would satisfy critics of the federal government's enforcement of the CWA.

Beyond the direct role of federal agencies, society benefits immeasurably when publicly funded research integrates objective scientifi c, economic, and technical analysis into policy-making decisions. Numerous agencies and programs within the Departments of Agriculture, Commerce, and Interior provided support for research, extension, technical assistance, stakeholder meetings, publications, and communications. The USDA's regional aquaculture centers funded numerous research projects since the late 1980s to characterize effl uents, improve environmental management, and understand environmental interac-tions. This includes an interregional initiative focused on coordinated research, education, and public policy projects. Numerous USDA agencies (Agricultural Research Service, Cooperative State Research, Education and Extension Service, Natural Resources Conservation Service, and the Risk Management Agency) contributed to national-, regional- and state-level BMP development projects. The NOAA National Sea Grant College Program was particularly supportive of regional initiatives to develop environ-mental codes of practice and BMPs for shellfish aquaculture and aquaculture producers in the Great Lakes Region. The USEPA administers numerous competitive grant programs and the National Center for Environmental Research has funded aquaculture projects (Jin et al. 2003).

State regulatory agencies

Although often not perceived as such, state regulators are also stakeholders in the policy-development process. Their principal objective is to address their mandate to protect and conserve natural aquatic resources, which may require stronger environmental protection than provided in the federal framework. State regulatory agencies are often at the forefront of federal regulatory enforcement. Very often, new federal regulations are developed without additional federal support for enforcement. Thus, state regulatory agencies may oppose the development of new federal regulations because they are seen as an unfunded mandate from the federal government that imposes a fi nancial burden on the budgets of state governments. Additionally, agency representatives from states with significant aqua-culture were familiar with the environmental performance and regulation of aquaculture within their state and believed that additional regulation was not warranted based on their experiences.

Many state regulators are strong advocates for BMPs as a regulatory option. Environmental management programs based on BMPs usually involve one or more state agencies with the authority to administer discharge regulations for aquaculture facilities. These regulatory agencies have numerous options available for regulating discharges from aquaculture facilities, and many choose to experiment with BMPs or codify them into state law. State agencies can provide in-kind contributions of technical expertise and assistance, guide applied research to create or improve environment management practices and waste management technologies, and in some cases, provide resources for cost-sharing BMP implementation.

University scientists

University scientists play an important role in the policy development process by provid-ing factual knowledge that can serve as the scientifi c underpinnings of environmental regulation. In this way, university scientists function more as information providers, rather than advocates of a particular policy position. In the AETF process, most university scientists that participated in the AETF process were affi liated with Land Grant institu-tions, which historically have strong research and extension partnership programs with agriculture, and, in this case, aquaculture interests. Thus, university scientists may be perceived by some representatives of environmental NGOs as not entirely objective because of this close association. However, university research and extension programs played critically important roles in organizational leadership, conflict resolution, commu-nications, applied research, scientifi c expertise, technical assistance, publications, and educational programs to support the development of BMPs at local, state, regional, and national levels. Land Grant, Sea Grant, and other university programs have made signifi cant contributions to developing successful BMP initiatives. University programs can contribute practical research and field demonstrations of BMPs. They also provide technical assistance to facility operators in adopting BMP plans or customizing BMPs to site-specific conditions.

Environmental nongovernmental organizations (NGOs)

Environmental nongovernmental organizations (NGOs) can be extraordinarily vocal and have fi nancial resources to mobilize their membership and the public at large to exert influence on public policy decisions (Tiersch and Hargreaves 2002). The role of environ-mental NGOs as contributing partners in developing BMPs has varied, depending on the intensity of opposition to a specific issue or aquaculture setting (e.g., salmon farming in net pens). Environmental NGOs were invited to participate in an open, solutions-oriented manner to identify potential and actual adverse environmental effects and offer reasonable recommendations. Several national environmental NGOs agreed to participate and submit-ted independent, formal comments to USEPA that identifi ed areas where the environmen-tal performance of aquaculture could be improved. Most wanted to be informed about planned AETF activities, but some hesitated to be recognized as named participants. Hesitation sprang from the perception that active participation might dilute their influence through alternative approaches and would suggest unqualifi ed endorsement for the AETF-sponsored process. There was some distrust based on the perception that the AETF favored industry positions. However, the AETF and environmental NGOs benefited from informa-tion exchanges at AETF public meetings. The experience of many state-level activities is exemplified by those in Maine and Florida, where BMPs were developed through stakeholder technical advisory committees consisting of agency-invited farmers, and representatives from the relevant state agencies, university extension services, and environmental NGOs.

Forms of Stakeholder Input

Active engagement of all stakeholders, even in the initial stages of the process, can expand the range of opinions and technical expertise available to develop effective environmental management programs. All interested parties should have ample opportunity to provide input throughout the public rule-making process. In the case of the USEPA national regulation, public input consisted of a mix of information, data, recommendations, and opinions from diverse stakeholders. To be effective, stakeholders must be willing to sustain active participation for the duration of policy development.

In recognizing the important role of stakeholder input, the motivation to support or oppose an issue varies widely. In some states, the decision to issue a permit or license requires local citizen input. This can lead to decisions influenced by parochial self-interest (e.g., an unobstructed ocean view). The intensity of opposition by local stakeholders can often deter aquaculture development, especially in populated coastal areas. A participatory stakeholder process must accommodate a variety of social dimensions during regulatory development.

As one example of the results of a process with broad stakeholder participation, the governor of Maine became a stakeholder through submission of a white paper that addressed the need for state-level policy decisions amid mixed stakeholder support for aquaculture (Apollonio 2004).

The Texas BMPs were developed by the Texas Commission on Environmental Quality as an outcome of administrative hearings and input from Texas Parks and Wildlife and a national environmental organization. The Missouri Aquaculture Council introduced the concept of adding BMPs to the Missouri general aquaculture permits. The council consisted of state agencies, industry, and university representatives and was cochaired by representatives from the Departments of Agriculture and Conservation. The Florida BMPs were initially developed by the public and academic technical advisory committees, reviewed internally by the lead agency, and then submitted to an administrative rule development process that included public comment and legislative review.

The development of new or revised environmental regulations by the USEPA under the CWA requires public review and comment. This also applies to the process of issuing new and renewed NPDES permits (Federal Register 2006). Although statutory public comment plays an important role in allowing open participation in policy making, soliciting com-ments is a passive procedure that may not provide adequate or representative input to assist with policy development.

Although open participation is essential to any policy-development process, stakehold-ers with entrenched ideologies or philosophies may become overly infl uential and science can become marginalized. The result, unless the input is balanced, can be unreasonable criteria or ad hoc approaches for reaching a critical decision. Stakeholders who have assumed a public identity but are motivated by self-interest can misrepresent the public interest because of a priori positions. The burden of proof may then, by default, become the responsibility of a business or industry potentially impacted by a regulation or proposed regulatory option (i.e., precautionary principle). Those affected by decisions must take positions, accept the outcome of scientifi c inquiry, and be fully engaged in supporting environmental protection and conservation through responsible actions to gain broad public support.

THE ROLE OF SCIENCE IN POLICY DEVELOPMENT

Throughout this chapter, terms such as *science, scientifi c, sound-science, science-based, best available science,* or *peer-review* have been used to signify information or data pro-duced by objective, testable, hypothesis-driven analysis of physical and biological phe-nomenon that has undergone independent evaluation by knowledgeable peers. Unfortunately, these phrases, which are highly valued in the scientifi c community, have been put to questionable use to achieve political goals (Mooney 2005). Misuse of these terms and other related issues have triggered thoughtful commentaries by fishery professionals (Lackey 2004; Hilborn 2006). The term *consensus science* implies a preponderance of scientifi c evidence for which there is wide or general agreement.

Even *best* available science can suggest selectivity or value judgment regarding what is best within the body of scientifi c information on a particular issue. In extreme cases, scientifi c information may not even be scientifi cally valid. Public trust in science is the greatest resource at stake in the dialogue about complex environmental or health risks. The best science might be judged by how well it considers the full scope of the relevant literature and the complete scope of the issue. It is a challenge to fully understand all the ingredients of a constructive dialogue on science, especially those that optimally generate public trust and effective policy (Luoma and Löfstedt 2007).

The process of regulatory development should use the best available scientifi c knowl-edge to guide and support decision making. Some stakeholders may consider current science inadequate or there may be limited scientifi c information to inform policy. Scientifi c uncertainties create difficulties for environmental policy makers. Agreement must be attained to accept scientifi c information that will lead a reasonable, objective person to reach a credible conclusion. This may present a dilemma if scientists disagree on the interpretation of the available information or the need for additional studies. The debate can become *whose* science is valid or *which* interpretations of complex studies or data should be accepted. An initial agreement on the ground rules for how science will be used in any policy-making process can be helpful, in addition to a mutually agreeable peer-review process to validate recommendations and gain broader trust.

Most environmental programs that assess new regulatory options include a review of current scientifi c and technical literature. Knowledgeable individuals familiar with the literature are most effective in accomplishing this task. The determination of state-of-the-art knowledge provides opportunity to identify critical information gaps on regulatory questions and determine relevance on specific issues. For some issues there may be insuf-fi cient information or lack of tested science. Much of the theoretical and applied research conducted in the United States was not designed in consultation with regulatory authorities to use approved sampling methodologies, sample chain-of-custody, laboratory procedures, quality controls, or other standards that are acceptable in a court of law under the legal framework of "good laboratory practices." This has little if anything to do with the quality of the science, but can result in challenges to the interpretation or acceptance of research fi ndings in the context of regulatory development.

Within a diverse mix of stakeholders engaged in public rule making, there is commonly wide disagreement on the analysis of environmental impacts and desired outcomes. During the early phases of rule making, knowledge levels can differ widely among stakeholders, who may have strongly held perceptions rather than facts. Technical contributions by recognized experts can facilitate a more comprehensive understanding of the issues among diverse stakeholders and allow the attainment of reasonable and mutually acceptable out-comes. Science can provide validated baselines of data and actual or projected environ-mental effects based on modifying variables in reliable simulation models. Scientifi cally developed and validated tools can help evaluate potential and actual environmental benefits.

For USEPA, the goal of using "sound science" is clearly stated in agency strategic plans. Numerous state agencies and BMP advocacy organizations also stress the need for science to develop effective and relevant BMPs. The AETF functioned under the principle of using the best available scientifi c knowledge to inform policy options. The role of science in federal programs related to aquaculture development has been reviewed from a historical perspective (Tiddens 1990; Jensen, in press). Value systems of society or advocacy orga-nizations and political management ideologies can be critical factors that determine the relative role of science in public policy making. Public benefits from federally funded science can be increased significantly when the resulting knowledge is used to formulate science-based policies.

5

Development, Implementation and Verification of Better Management Practices for Aquaculture

INTRODUCTION

The purpose of better management practices (BMPs) for aquaculture is to provide an on-farm management system designed to minimize adverse environmental and social impacts or assure food safety. BMPs, as an approach to achieve environmental protection or con-servation goals and an alternative to prescriptive regulations, have been promoted by public agencies and farm organizations in the United States since the late 1990s.

Although the impetus for creating BMPs has ranged from altruistic self-governance to legislative mandate, the process to develop, implement, or verify BMPs has many com-monalities. Presented here are the cumulative experiences and observations of those activ-ities that have resulted in workable and effective programs, which continue to evolve.

Aquaculture affects a diverse group of stakeholders that includes producers; input suppliers; fish processors; seafood buyers; consumers; government agencies; members of affected communities; and environmental, social, and consumer advocates. Representatives of each group should be involved in the process of formulating BMPs. It is essential to reach a consensus among stakeholders about the issues to be addressed by BMPs. Some stakeholder concerns might seem irrelevant to scientists and other technical experts, but all concerns must be openly discussed among stakeholders to create a credible and trans-parent process that will lead to an effective suite of practices that will be acceptable to all affected parties.

Once the issues of concern have been agreed on, technical experts should be given responsibility by the stakeholder group to prepare draft BMPs that address each issue. Of course, the entire range of stakeholders, and especially producers, should be involved in revising draft BMPs. It is essential for the BMPs to address all major issues, but they also must be practical, implementable, effective, and acceptable to the stakeholders.

The rate of adoption of BMPs and the degree of benefit accrued from their use should be documented. Of course, compliance with governmentally mandated BMPs must be verifi ed by the appropriate agency. Adoption of BMPs in a certifi cation program requires independent, third-party verifi cation.

The purpose of this chapter is to provide an overview of the general process that should be followed in developing, implementing, and verifying BMP programs for aquaculture. This process should be viewed as continuous and iterative. Based on the results of the monitoring program to verify performance, BMPs can be adjusted or new BMPs developed for implementation.

USES FOR BMP PROGRAMS

A program based on BMPs offers substantial fl exibility to achieve environmental protec-tion goals. The BMP approach does not dictate specific methods, but offers a range of options that can be selected and adapted by producers to site-specific conditions. Adoption of BMP programs can be voluntary or mandatory. Voluntary programs are intended to provide environmental protection while enhancing market opportunities for producers. Mandatory programs are used by governments to achieve environmental protection as an alternative to prescriptive numerical limits.

Voluntary Adoption

Voluntary adoption of BMPs is the most desirable form of implementation because it sig-nifi es that producers understand the importance of responsible aquaculture and are willing to expend the effort and bear the costs necessary to attain it. However, in reality, voluntary adoption of BMPs probably is the most uncertain method of implementation. There are several reasons for this. Producers may assess their operation and simply declare that their farms already are in compliance with the better practices. For example, the Global Aquaculture Alliance (GAA) developed a series of BMPs for voluntary adoption by shrimp farms. A self-graded evaluation sheet with a scoring system was provided. Many producers self-graded at 90% or higher out of a possible score of 100%. A similar response indicat-ing high levels of BMP implementation was received from catfish producers in Alabama. The senior author also heard several individuals from Alabama and other states declare that essentially all catfish producers were in compliance with the BMPs. In both examples, there were few, if any, producers who were actually in total compliance with the BMPs.

Investment and credit screens

It is not uncommon for principals in proposed or existing aquaculture projects to seek investors or loans. Commitment to implementation of BMPs would demonstrate environ-mental and social stewardship and should be attractive to cautious investors or lenders. Studies showing that BMPs are cost effective would be particularly valuable to those considering investments or loans to aquaculture facilities. The International Finance Corporation (IFC) often makes loans to large aquaculture projects in developing countries. The IFC asks their clients to comply with some environmental regulations. In the future, adoption of BMPs may become a common requirement for farms seeking investors or loans.

Use by seafood buyers

Many seafood buyers are interested in providing products with a history of responsible production. The Label Rouge (Red Label) program in France is one example. Although this program emphasizes product quality and freshness, it also has some environmental requirements with which the producer must comply. Several large supermarket (e.g., Whole Foods, Wal-Mart) and restaurant chains (e.g., Darden Foods/Red Lobster) seek seafood products that are produced with "environmentally friendly" and "socially respon-sible" techniques.

A few of these organizations have formulated specifications for their producers and hired employees to source products from responsible producers.

Consumers have grown increasingly concerned about seafood safety, especially the accumulation of mercury and other persistent bioaccumulative toxins, antibiotic residues, and other potentially harmful substances in seafood. Producers who implement BMPs are not likely to be sources of contaminated produce. Such producers are heavily favored by buyers.

Environmental sustainability certifi cation programs

The Global Aquaculture Alliance (GAA) has developed certifi cation standards for shrimp aquaculture and licensed these standards to the Aquaculture Certifi cation Council (ACC). The GAA currently is developing standards for fish aquaculture certifi cation. The World Wildlife Fund (WWF) is developing certifi cation standards for the most common aqua-culture species in international trade (Boyd et al. 2005) and Environmental Defense is developing shrimp aquaculture standards. These standards have specific numerical criteria and record-keeping requirements. For example, the GAA has an effl uent standard with water quality limitations for pH, dissolved oxygen, biochemical oxygen demand, total suspended solids, ammonia nitrogen, total phosphorus, and chloride. At the farm level, the most feasible way of obtaining compliance with the standards is through the adoption of BMPs. Of course, each farm is different, and the actual combination of BMPs necessary for compliance with certifi cation standards will differ among farms.

Mandatory Adoption

Several states and industry organizations have taken steps to develop voluntary BMP programs. These BMP programs are voluntary in principle, but local regulators can make them mandatory for facilities that are currently exempt from the national effl uent limitation guideline because of system type or annual production level. This situation puts pressure on facilities that use voluntary environmental codes of practice that include BMPs to demonstrate BMP program effectiveness as an option over mandated requirements.

Government regulations

Governments develop various regulations applicable to aquaculture to include restrictions on siting, exotic species, antibiotic and chemical use, wildlife depredation deterrence, water use, effl uent composition and volume, and possibly others. Compliance with these regulations obviously will affect the range of practices that can be used at an aquaculture facility and can, potentially, increase the costs of production. For example, if a particular nuisance species of bird or marine mammal cannot be killed, a variety of more costly nonlethal techniques may be required. Laws to protect wetlands will eliminate the practice of building farms partially or totally in areas delineated as wetlands. This may limit farm size, require the acquisition of additional land for the construction of compensatory wet-lands, or require the purchase of more expensive land.

Water pollution control is a major goal of many governments, and discharge permits may be required for aquaculture. The new aquaculture effl uent rule in the United States (Federal Register 2004) implemented by the United States Environmental Protection Agency (USEPA) is a good example. This rule requires National Pollutant Discharge Elimination System (NPDES) permits for concentrated aquatic animal production facili-ties, but it does not specify effl uent limitation guidelines. This leaves individual NPDES-delegated states free to determine the conditions required in NPDES permits for aquaculture. In some cases, states will likely require implementation of the BMPs, or similar BMPs, as described by the USEPA (see Chapter 4). Other states may specify certain limitations on effl uent water quality concentrations or a combination of effl uent water quality criteria and BMPs. Even where BMPs are not specifically required, produc-ers will need to implement better practices in order to comply with water quality restric-tions for discharges.

One long-term concern is that resource-strapped government agencies will indiscrimi-nately adopt sets of BMPs developed for other locales or culture conditions, and make them mandatory without regard to applicability. The model of "one size fi ts all" is too infl exible for unique facility conditions and sites; BMPs may be available but not neces-sarily relevant to local conditions. The BMPs may be practices that a regulator believes a farmer should follow irrespective of the availability of data demonstrating that an environmental benefit will be realized or that implementation and maintenance costs are reasonable. Real-world experience may yield unexpected results that can be remedied by revising a BMP, but regulatory programs must be sufficiently fl exible to accept change.

DEVELOPMENT OF BETTER MANAGEMENT PRACTICES

There has been much effort toward developing documents containing BMPs for a number of aquaculture species (see Chapter 4) and especially for marine shrimp (Boyd 2003a), rainbow trout (*Oncorhynchus mykiss*) (MacMillan et al. 2003), and Atlantic salmon (*Salmo salar*) (Brooks et al. 2002). Individual farms and local producer groups also have become interested in improving management practices. For example, the Alabama Catfish Producers (ACP), a group of mainly channel catfish producers in a fi ve-county area of west-central Alabama, followed a formal procedure to develop BMPs (Boyd et al. 2003). The ACP is cooperating with Auburn University, the United States Department of Agriculture Natural Resources Conservation Service (NRCS), and the Alabama Department of Environmental Management in developing aquaculture effl uent regulations based on BMPs (Boyd and Hulcher 2001). Moreover, the ACP is currently working with WWF to use the BMPs as the basis for an environmental certifi cation program for channel catfish. Supermarket chains and other large purchasers of seafood have expressed interest in aquaculture species produced by good practices. In response to this potential market, WWF, the ACC, and Environmental Defense in the United States, and several European organizations are con-sidering programs for certifi cation of several aquaculture species.

Two other notable examples are the development of a Code of Conduct by the Maine Aquaculture Association that was reviewed by members and external reviewers, including the United Nations Food and Agricultural Organization (FAO) and WWF. Signifi cant portions of the Code, over time, have been adopted into state regulations with a manage-ment committee consisting of industry, federal and state agencies, and nongovernmental organizations to prevent, control, and manage the escape of farm-raised salmon.

The Pacifi c Coast Shellfish Growers Association, with the input of growers from eight self-described shellfish regions in Alaska, California, Oregon, and Washington, has engaged in an environmental certifi cation process. This has occurred through a stepwise effort that initially created a code of conduct followed by BMPs. In turn, this has led to planning for an environmental certifi cation process managed through a market-chain sup-ported organization. (See Chapter 4 for a complete description of the Alabama Catfish Producers, Maine Aquaculture Association, and Pacifi c Coast Shellfish Growers Association efforts.) There is likely to be a proliferation of efforts to develop BMPs and possibly certifi cation programs for many aquaculture species.

Agency-driven efforts to develop BMPs as an environmental protection or conservation effort are public processes that require, to a greater or lesser degree depending upon federal or state laws, direct stakeholder involvement. Stakeholder involvement by state agencies in the United States has generally been organized through the creation of technical advi-sory committees with membership comprised of commercial producers, aquaculture exten-sion specialists, agency representatives, and environmental organizations. They begin their work by either debating and refi ning a state agency draft document or creating an original draft through a collaborative effort. An innovative risk analysis process was used to develop agency-adopted BMPs to mitigate potential risks (i.e., genetic, ecological, and disease) of culturing exotic species (Zajicek and Metcalf 2002).

If fi nalized BMPs are to become regulation, a formal review process for congruence with constitutional guidelines and existing laws will occur internally within the lead agency followed by an announcement to elicit public comment. Regulatory development can be lengthy and time consuming, with controversial provisions triggering opposition by agencies other than the lead agency, or public opposition that will play out in public workshops or hearings and may, if unresolved, result in a challenge through an administra-tive hearing process for resolution by an impartial hearing offi cer.

Increasingly, many states in the United States are adopting a BMP approach to envi-ronmental protection because of generally effective results and an inherent fl exibility to encompass small and large operations. This refl ects the experience that the development of aquaculture-related BMPs in the United States has not resulted in confrontational legal action or interagency "turf" battles. BMPs created and adopted into regulation in Hawaii, Florida, Louisiana, New Jersey, and other states (described in Chapter 4) are examples of the outcomes of these processes.

The suggested process for developing BMPs is summarized in a flowchart (Fig. 5.1) and described in more detail below.

Environmental Survey

The initial step in developing BMPs for production of a particular species in a given area is to conduct a survey of production activities and their potential effects on the environ-ment and nearby communities. There may be so many producers that a sample of farms must be selected for the study. The sample should be a random selection of farms to assure that it is broadly representative of the production activity. The survey probably should include farms accounting for 10 to 25% of total production within the area of interest.

The survey should be conducted by a small team of individuals with collective expertise in aquaculture, environmental science, and social science. These individuals should be responsible for designing the survey instrument for capturing the information needs listed in Table 5.1. The return on surveys through postal or electronic mail is low. Therefore, the investigators should visit each farm and use direct interviews with producers to fill in the survey instrument.

Environmental Assessment

The fi ndings of the environmental survey should be summarized in a report because the information will be used in later stakeholder meetings. The report should describe the aquaculture sector and how it has been superimposed on landscapes and communities. Production facilities and methods should be described. Published research fi ndings and extension recommendations on feed use efficiency, aeration, effl uents, and other aspects of production with potential environmental impacts should be summarized. A careful description of water use within the culture facilities is necessary. There should be thought-ful analysis of resource availability, competing uses, and possible environmental impacts in which wasteful and environmentally harmful aquacultural practices are identifi ed. In addition, the possibility of negative social aspects of farms should be ascertained. Investigators should prepare lists of BMPs that could be useful in preventing wasteful use of resources and each negative environmental and social impact. The BMPs also should include practices for preventing contamination of aquaculture products at the farm level.

The report should be sent to three to fi ve external reviewers (not local stakeholders) and the reviewers' comments incorporated in the fi nal draft. Examples of reports on the envi-ronmental status of aquaculture industries are available for marine shrimp farming in Thailand (Tookwinas 1996) and channel catfish farming in Alabama (Boyd et al. 2000).

Stakeholder Meetings

The credibility and effectiveness of BMPs are increased greatly by the involvement of a variety of stakeholders. These stakeholders should include producers, local extension specialists, local environmentalists, regulatory agencies, and representatives of communi-ties situated near farms or facilities. In addition, the team responsible for the environmen-tal survey and report, aquaculture specialists, representatives of international NGOs, offi cials from processing plants, and possibly others should be invited. Typically, the most contentious exchanges in multi-stakeholder meetings have occurred between environmen-talists and industry representatives concerning the environmental and social impacts of aquaculture.

These debates are characterized by polarized arguments pertaining to the types and magnitude of the environmental impacts of aquaculture, resulting in extreme views and little compromise. It is interesting to note that, within the aquaculture com-munity, some of the staunchest opponents of the claims of environmentalists are trying to address environmental impacts through BMPs and codes of conduct. In doing so, the interests of the aquaculture community and environmentalists are synchronized; thus, it is quite sensible to involve environmentalists more formally in BMP development processes.

The number of participants can be limited according to the judgment of the organizers, but 20 to 25 participants would be a typical group. In some cases, a smaller subset of the main stakeholder group can serve as a technical advisory committee for initial BMP development. Restricting participation in stakeholder meetings is very difficult because many stakeholders have a vested interest in the BMPs being developed. Thus, meeting planners and facilitators should have a good, general understanding of the individuals and groups who would be most affected and interested in the particular type of aquaculture being discussed. In many cases, consultants and professional facilitators are hired to do this specific job. Nevertheless, the issues should be thoroughly deliberated because exclu-sion of a concerned party may be detrimental to the credibility of the BMP program in the future.

The discussion at the fi rst stakeholder meeting, in which all representative stakeholders are present, should be reserved for scientifi c considerations on the identifi cation of the most important issues. It is helpful to maintain a workshop setting and record concerns as they are articulated to provide some assurance to stakeholders that their concerns are being legitimately considered. A certain level of trust needs to be established among participants of multi-stakeholder meetings, and if this is achieved sufficiently early in the process, a sense of teamwork and unity-of-purpose often develops.

It is unlikely that mutually agreeable BMPs can be developed in a single meeting. Committees should be given assignments related to the process, and these committees could work separately to prepare material for discussion by all stakeholders in fi nalizing BMPs.

Guidelines for Writing BMPs

In developing BMPs, it is most convenient to consider several categories of potential environmental impacts and provide a range of BMP options to address a particular impact under each category. Producers could then select the BMPs appropriate for individual sites and operations. For example, Boyd and Hargreaves (2004) suggested eight categories of BMPs for channel catfish aquaculture: site selection and pond construction, liming and fertilization, feeds and feeding, solids management and disposal, use of drugs and chem-icals, mortality removal and disposal, management of escapees, and general facility oper-ation and maintenance. However, the Alabama catfish farming BMP effort (Boyd et al. 2003) required 15 categories in order to satisfy all stakeholders. Several of these categories were added specifically at the request of the state agency with responsibility for environ-mental protection.

The environmental assessment report provided to stakeholders should have BMPs arranged by categories. However, this presentation of BMPs should be considered pre-liminary and the basis for further discussions that will lead to the fi nal BMPs. The catego-ries can be altered according to the judgment of the stakeholders, and individual BMPs can be selected or rejected based upon the opinions and experience of the group.

Previous studies about the effects of various treatments and management strategies for the particular aquaculture activity should be considered in selecting appropriate BMPs. The input of producers is especially important because they can provide opinions on the degree of difficulty and costs of implementing individual practices. On occasion, estab-lished producers may suggest BMPs to a BMP technical advisory committee that would be difficult or costly for new producers to adopt and implement. It is unclear whether these BMPs are innocently proposed based on experience gained with production systems or species or as an indirect means to deter competition, but the involvement of objective aquaculture extension specialists and knowledgeable agency representatives can serve as a way to assess the value of such proposals.

The goal of BMPs is to reduce environmental and social impacts, and where possible, estimates of benefits should be quantifi ed, including estimates of performance variation and uncertainty. For example, calculations of reduction of nutrient loads in effl uents result-ing from improving feed conversion through the use of better feed management can be calculated. The likely effect of sedimentation basins on effl uent quality also can be cal-culated. Many stakeholders will want assurances that the proposed practices will have positive effects on the environment; thus, there should be a scientifi c basis for effective-ness or experience from previous applications of practices in other kinds of aquaculture, in agriculture, or in unrelated industries to support effi cacy. Examination of BMPs recommended in other types of aquaculture can be especially useful. Boyd (2003b) discussed farm-level issues in environmental management of aqua-culture effl uents. He emphasized the need to categorize BMPs and gave an example of BMPs for preventing erosion of pond watersheds, embankments, bottoms, and discharge canals that may be a significant source of suspended soil particles in effl uents. Boyd (2003b) also provided a list of BMPs that could be adopted to minimize nutrient loads in aquaculture effl uents and reduce the likelihood of eutrophication in receiving water bodies.

Presentation of BMPs

The fi nal BMP document should explain the environmental, social, or conservation problem that each category of BMPs will address and provide either guidance to producers for implementing practices or a directed educational effort by the responsible agency or active partnership with aquaculture extension programs (Box 5.1). The fi nal draft BMPs should be placed on a website and published as a manual.

IMPLEMENTATION OF BMPS

Compared to BMP implementation, the task of developing BMPs is relatively simple because the major negative environmental impacts and the methods for preventing these impacts are easily identifi ed. The main difficulty is obtaining agreement among stakeholders about issues where differences of opinion arise about the costs and benefits of chang-ing or prohibiting longstanding production practices. The implementation of BMPs is a much more challenging task. Many producers are asked to alter production practices according to guidelines provided in a document prepared by selected stakeholders, and producers may be reluctant to expend the time, effort, and money necessary to implement BMPs correctly. Voluntary adoption of BMPs depends on the extent to which those BMPs affect farm profitability.

Nussbaum et al. (2003) describe a process of phased implementation of BMPs for for-estry that could be adapted for aquaculture. The forestry BMPs are implemented in generic modules that address legal, technical, environmental, and social issues. In the phased approach,

BMPs that achieve a minimum or baseline level of performance are addressed fi rst. The implementation process has three components: 1) an initial review to assess the current situation; 2) development of an action plan to improve practices, including a time-table; and 3) implementation of the action plan.

Programs to Facilitate Implementation

The implementation of BMPs can be a technologically complex process. Thus, programs that assist producers with implementation of BMPs are valuable. Organizations preparing BMPs for voluntary adoption, use in governmental regulations, or use in certifi cation programs should dedicate resources to programs that facilitate and encourage adoption of BMPs. These include a range of outreach and education programs, economic incentives, cost-benefit analysis, and applied research.

Broad acceptance and adoption of BMP programs beyond regulatory requirements require incentives and leadership by industry organizations in partnership with regulators. There are few states or industry organizations with dedicated fi nancial resources to accom-plish outreach activities. States that lack strong industry associations or extension educa-tion programs will have difficulty providing outreach activities to communicate benefits of the BMP approach. States that lack resources for monitoring, verifi cation, and enforce-ment are vulnerable if problems occur with the BMP approach to improve environmental quality. Partnerships at the local level with state governments, federal assistance programs, and university extension programs can play critical roles to leverage limited resources to support BMP implementation.

Education programs

Educational and outreach activities are essential to promote adoption of innovative tech-nology and practices at aquaculture facilities. Where possible, workshops should be con-vened to provide technical assistance to producers interested in implementing BMPs. Owners or managers of large farms may hire private consultants to help install BMPs and conduct monitoring, but small farms need government or university education programs to provide on-site technical assistance.

Producer meetings can be used to explain the benefits of BMPs and provide advice on their adoption. For example, southeast Massachusetts shellfish farmers were introduced to BMPs through meetings and announcements linked to state and federal incentive programs for BMP implementation. The Texas Commission on Environmental Quality presented permit and BMP information during aquaculture association meetings and through the permitting process. Maine farmers learn about BMPs and their implementation through Maine Aquaculture Association (MAA) activities and the state agency permit and leasing processes. The education and outreach activities of the Maine Aquaculture Association are funded through membership fees. The New Jersey Department of Agriculture includes farmer education as a component of compliance site visits.

Many entities have developed collections of voluntary BMPs for aquaculture, but few include specific details on how to install BMPs. Written instructions or detailed technical guidance manuals should be part of any BMP education program (Box 5.2). Examples include the *Idaho Waste Management Guidelines for Aquaculture Operations* developed by the Idaho Division of Environmental Quality through the efforts of a stakeholder advi-sory committee (IDEQ, undated). This technical manual is available on the Internet (www. deq.idaho.gov/water/prog_issues/waste_water/pollutant_trading/aquaculture_guidelines. pdf) to assist producers to develop a waste management plan composed of effective BMPs suitable for the farm location, size, and design. In addition, the University of Wisconsin Sea Grant Institute has published a manual entitled *Best Management Practices for Aquaculture in Wisconsin and the Great Lakes Region* as a fl exible document that can be modifi ed for each state in the region with the regulatory requirements for that state (Malison and Hartleb 2005). The Pacifi c Coast Shellfish Growers Association (see Chapter 4) has provided specific guidance for BMP installation. Installation of BMPs that involve infrastructural or management changes may be complex. For example, a feed management BMP may state that feed should be applied conservatively to avoid overfeeding and uneaten feed. Many producers may not know the most widely agreed-upon procedures for preventing overfeeding, and they would not apply the best available practices. Clear instructions on methods of installing infrastructural and management changes necessary to comply with BMPs are a necessary part of implementation. As an example of an integrated educational effort, the Florida Department of Agriculture and Consumer Services (FDACS) partnered with the University of Florida Department of Fisheries and Aquatic Sciences to hold producer meetings and workshops. Explanatory signs that illustrate and describe BMPs have been installed at four university aquaculture demonstration and research facilities (shellfish, food and bait fish, shrimp, and ornamental fish and aquatic plants). The FDACS also publishes a free newsletter to notify farmers, agencies, legislators, and other interested parties about BMP developments or revisions.

Economic incentives

Together with education programs, economic incentives can promote adoption of BMPs, particularly those for which the incentive of improved profitability is not present. Traditionally, these take the form of cost-share programs, direct subsidy payments, or tax benefits. In the United States, there are no dedicated public funds available to assist aqua-culture producers with programs to adopt, implement, or construct BMPs that require equipment, structures, or farm remodeling. As a notable exception, shellfish producers in Massachusetts successfully applied for state funds and funding from the Environmental Quality Incentives Program (EQIP), a program administered through the NRCS of the United States Department of Agriculture. The funds were used to implement a shellfish Conservation Practice Standard that was adopted by NRCS as a result of producer efforts to create shellfish farming BMPs.

Cost-benefit analysis

Demonstrating economic benefits can stimulate widespread adoption of BMPs. Boyd and Tucker (1995) and Engle and Valderrama (2002) evaluated production practices used in channel catfish farming in the United States and concluded that 1) the better practices appeared to be the most sustainable and profitable and 2) BMP adoption is most success-ful where there are strong economic incentives to do so. Studies that demonstrate the cost effectiveness of BMPs would be a great help for inducing producers to voluntarily adopt better practices. Development, implementation, and verification of BMP facility plans represent a cost in money, labor, and physical resources that may or may not result in measurable economic benefits to producers from improved operational efficiency or improved environmental quality. In any case, these costs can be more easily absorbed by larger operations that realize greater effi ciencies through economies of scale. Producers cannot pass increased production costs on to processors and distributors.

Implementation costs can be burdensome and, in extreme cases, force closure of small businesses. However, BMP programs can minimize the loss of small farms because time and costs can be reasonable compared to overly prescriptive regulatory options. BMPs provide fl exibility, with more options across a range of farm sizes, to achieve desired environmental protection outcomes.

Some BMPs are easier or less expensive to implement than others. A BMP for conserv-ing water and reducing overflow after rainfall requires pond managers to maintain storage volume in ponds by leaving water levels about 15 cm below the overflow elevation when replacing seepage and evaporation. This BMP costs nothing and is obviously effective and simple. On the other hand, considerable effort and cost will be required to plant grass or provide stone cover around the edges of a pond to prevent erosion. Thus, producers may inplement those BMPs that seem convenient and omit the others. Partial implementation of BMPs is better than no implementation, but great improvement cannot be expected from such a procedure.

As a regulatory option, BMP programs can protect natural resources, maximize envi-ronmentally beneficial outcomes, minimize impacts on neighbors, and reduce regulatory costs and burdens. The formation of industry-academic-regulatory partnerships or other coalitions among industry organizations can assure that BMPs and regulatory requirements address real environmental concerns rather than impose burdensome regulations with limited environmental benefit. Programs to monitor and document environmental benefits and costs of voluntary BMP programs will help justify this approach as a regulatory option. Lack of quantitative data may create questions about the effectiveness of BMPs, thereby justifying other, more rigid regulatory options such as end-of-pipe numerical limits.

Applied research

Adequate research and demonstration infrastructure is needed to support the dynamic progression of improved practices and adoption of innovations for environmental manage-ment. The diversity of aquaculture species, production systems, and geographic locations creates a matrix of choices and potential environmental effects that require science and applied knowledge. BMPs for new species and systems are likely to evolve in the future. There is a need to build upon applied research to determine environmental effects of aquaculture and to create, test, and improve BMPs for aquaculture.

New scientifi c tools are creating knowledge about aquatic ecosystems that may gener-ate new issues and concerns. The protection of natural aquatic resources is a benefit to the general public, so publicly funded research plays a significant role in developing BMPs. The public also gains benefit from this investment when new information is integrated into the development of public environmental policies. Aquaculture likewise benefits from research discoveries that protect natural aquatic resources. As states continue to assess the quality of public waters, more water bodies are likely to be categorized as impaired, and much stricter end-of-pipe total maximum daily load limits may be applied to specific pol-lutants. Water use restrictions, legal access rights, and water availability are becoming critical issues in some locations as well. These situations will require new BMPs and technologies to conserve, reuse, or recycle culture water.

Many states recognize the need to revise and update BMPs based on new knowledge. The BMPs should be fl exible and revised based on regulatory, judicial, operational, scientifi c, and environmental developments. Two states have repeatedly revised their aquaculture BMPs and created new ones. Florida has produced four editions of their BMP manual. The Maine Aquaculture Association code of conduct is the source for many BMPs that are now components of NPDES permits for marine net pens. The Association continually revises their finfish BMPs and is creating BMPs for shellfish biosecurity and aquatic animal health. The United States Army Corps of Engineers is considering the PCSGA environmental codes, and growers' individual farm plans, as components of a general nationwide permit the Corps will issue for shellfish culture activities in navigable waters of the United States. Information to support the effectiveness of some BMPs may be lacking, and it may be necessary, as was the situation in developing BMPs for Alabama catfish farming, to conduct special studies.

For example, studies of sedimentation (Ozbay and Boyd 2003, 2004), persistence of copper residues (McNevin and Boyd 2004), effects of sodium chlo-ride treatment (Tavares and Boyd 2003), estimation of water budgets and effl uent pollution loads (Boyd et al. 2000), and impacts of farm effl uents on streamwater quality (Silapajarn and Boyd 2005) were made to support the selection of BMPs for Alabama catfish farming. BMPs will also need to be refi ned for specific species, production systems, or soil types, as proven in Arkansas for baitfish effl uent treatment using vegetated ditches (Frimpong et al. 2004).

VERIFI CATION OF BMP EFFECTIVENESS

The most difficult aspect of a program to improve the environmental and social perfor-mance of aquaculture likely is demonstration of the program's effectiveness. One objective of verifi cation is to confi rm that producers have implemented BMPs or that they are in compliance with governmental regulations or certifi cation standards. Even more impor-tantly, the objective of verifi cation is to determine whether compliance with the program is providing effective environmental protection. If it is not, the BMPs, standards, and regulations are not achieving their desired effect and should be improved. Verifi cation is accomplished through a compliance monitoring program, site inspections, or audits of facility records. In forestry, BMP effectiveness has been assessed using literature reviews, the number of citizen complaints, the resources committed to BMP implementation, inter-disciplinary field surveys, direct monitoring, aquatic habitat measures, and modeling (Ice et al. 2004).

Forms of Verification

Verification of BMPs can be accomplished by producers (fi rst-party verifi cation), an external organization with a relationship to producers (second-party verification), or an organization that is independent of producers or the standards-development organization (third-party verifi cation) (Wessels et al. 2001).

Internal verification

Producers can monitor their operations to determine and self-declare that BMPs are having the desired effect. Standards of compliance can be developed by an outside group or by the farm or facility itself. Self-evaluation, or *fi rst-party verifi cation*, can be highly effective if a producer is willing to make changes to practices if desirable and intended results are not being achieved. However, producers who voluntarily adopt BMPs often do not prop-erly install the improved practices and are unlikely to be critical of their own efforts.

External verification

Verification of compliance by an external body with some relationship with the farm or facility being certifi ed is called *second-party verifi cation*. For example, an industry asso-ciation can create standards and support compliance monitoring and BMP performance verifi cation of its members through inspections and audits.

Independent verification

Ideally, compliance with standards and requirements of lending agencies and certifi cation standards should be verifi ed by *third-party inspection*. The organization that sets perfor-mance standards should be independent of producers and marketers. In a certifi cation program, the inspector should be completely independent of the producer and should not be directly employed by the certifying organization. The inspector can be from an accred-ited certifying agency, a private consulting fi rm, or independent agents. The certifying organization would be required to train the inspectors and specify their assignments. The producer should pay an inspection fee to the certifying body and this body would pay the inspector. Furthermore, the certifying organization should be independent of the group that develops the standards.

COMPLIANCE MONITORING AND EFFECTIVENESS EVALUATION

Successful best management practices must be economically and socially beneficial, easily adapted and adopted, and environmentally effective (Watson et al. 1994). A comprehen-sive BMP program is an iterative, adaptive, and continuous process of BMP development, implementation, and evaluation. The success of BMP programs is related to educational efforts and subsidies that support implementation and monitoring to assess effectiveness. Compliance monitoring programs are used to estimate BMP implementation rates, refi ne BMPs, or target technical assistance programs (Prud'homme and Greis 2002). Although many BMPs have been proposed and developed for aquaculture, the effectiveness of most of these improved practices has not been evaluated formally through controlled research studies or informally through surveys or site inspections.

The experiences of BMP programs developed to evaluate the effectiveness of BMP programs designed to control nonpoint source pollution from forestry, agriculture, and stormwater runoff are illustrative to aquaculture. The performance of BMPs can be assessed by measures of effectiveness, efficiency, and performance (Strecker et al. 2001). Effectiveness is a measure of the extent to which a particular BMP or BMP system (i.e., integrated set of BMPs) meets stated improvement goals. Effi ciency is a measure of the extent to which BMPs remove particular pollutants or mitigate adverse effects. Performance is a combined measure of effectiveness and efficiency that quantifi es the extent to which a particular BMP or BMP system meets a stated environmental improvement goal relative to the maximum potential reduction at a particular location.

The main component of effectiveness evaluation is a program of compliance or imple-mentation monitoring. This monitoring program can include inspections or facility audits that result in a certifi cation of compliance. Nussbaum et al. (2003) describe a process of verifying BMP implementation that consists of three main activities. First, a baseline assessment of the farm or facility should be conducted. This assessment will serve as the basis for comparison with the results of future assessments. Second, the monitoring program should confi rm some minimum level of improved performance. Third, the mon-itoring program should confi rm a trajectory of improved environmental performance.

Studies that evaluate the effectiveness of BMPs in other domains have been hampered by the lack of consistent methodological design and standardized reporting protocols (Strecker et al. 2001; Prud'homme and Greis 2002). One attempt to standardize report-ing is represented by the online database of stormwater BMP effectiveness (www.bmpdatabase.org). This may be a good model to collate the experiences with imple-mentation of aquaculture BMPs.

Questions of appropriate scale (e.g., pond, farm/facility, watershed/basin, region) and variation in sampling techniques limit the utility of many fi ndings. Monitoring programs are more rigorous than site surveys because they assess specific and quantifi able chemical (e.g., water quality) or biotic impacts. Some programs use bioassessment to evaluate in-stream effects and report results in terms of various habitat quality indexes. Other pro-grams assess BMP effectiveness with comparative studies conducted before and after implementation, upstream and downstream of the BMP, or input-output studies conducted in paired watersheds. Compliance surveys or visual quality assessments to gauge the level of implementation are another option, but these are much more qualitative than formal monitoring programs because they simply indicate whether a particular practice is being implemented and the number of practices implemented. Various scoring methods can be used to gauge the implementation of BMPs, including presence/absence, pass/fail, and degree of compliance as a proportion of full implementation (Husak et al. 2005).

Compliance Monitoring in Aquaculture

Implementation of BMPs and compliance with governmental regulation or certifi cation standards will not always prevent negative environmental effects. Confi rmation of effec-tiveness must be found through environmental and social monitoring programs that include evaluation of off-site effects. Such programs have been rare in aquaculture because neither governments nor the private sector are often willing to bear the costs. A shrimp farm in Madagascar funded a study to determine the effects of farm effl uents on water quality in the receiving bay (McNevin 2004). A tilapia farm with cage culture operations in lakes in Honduras and Indonesia sought the senior author's assistance to initiate a program to monitor water quality in the lakes. The Alabama Catfish Producers provided funds to determine the effect of catfish farm effl uents on stream flow and water quality (Silapajarn and Boyd 2005). Some government agencies in several countries have conducted limited water quality monitoring in coastal areas with extensive areas of aquaculture.

The State of Florida employs a staff of four BMP compliance offi cers that have Master of Science degrees in fisheries, aquaculture, biology, or environmental sciences and an annual performance goal of at least two unannounced, on-farm compliance visits annually. Discharge permits for aquaculture usually require reporting according to a specifi ed time-line. In some cases, producers are allowed to make the analyses, but more often, a consult-ing fi rm or laboratory approved by the government must make the effl uent analyses. Verifi cation of compliance with mandated BMPs may be done unannounced by a govern-ment offi cial or by a certifi ed professional approved by the government. An in-depth discussion of BMP verifi cation by a variety of state governments in the United States is provided in Chapter 4.

In many countries, the government will make rules about aquaculture, but not have adequate personnel to inspect the facilities. Producers will know whether this is the situ-ation, and if it is, they will tend to ignore the regulations. Even in developed nations such as the United States, facilities may not be checked for compliance with some environmen-tal regulations unless the public makes a complaint. In some nations, corruption is common, and environmental regulations are mainly a means for governmental offi cials to obtain bribes.

Guidelines for conducting monitoring studies should be developed and more effort devoted to this important issue. However, such studies often are difficult to interpret because background data are seldom available, aquaculture is not the only source of pol-lution, and the expense to monitor certain production systems and environments (e.g., open ocean net pens) may hinder the development and implementation of practical and cost-effective monitoring (Alston et al. 2006).

Record Keeping

Record keeping is necessary even for voluntary BMP programs. However, they are most critical for discharge permits with numerical limits and for certifi cation programs. Governments usually require monthly or quarterly reports of concentrations of water quality variables measured to verify compliance with discharge permits, chemical and drug use, wildlife control, aquatic animal health certifi cations, and sales records for restricted species. Such record keeping is relatively simple compared to that required by certifi cation programs. These programs require records of the sources and numbers of animals stocked in ponds; amounts of fertilizers, feeds, and other chemicals added; and the disposition of the harvest.

Certifi ed products must have a bar code on each package that will allow them to be traced back to the pond or other culture unit of origin. The farm records must be sufficient to determine the origin of the stock and all of the substances used during culture.

Large farms are able to assign the responsibility for record keeping to one or more specific individuals. Small, family-operated farms may fi nd record keeping to be a major effort.

Review and Correction for Noncompliance

Environmental management systems always include the requirement for periodic review and correction for noncompliance. The same should be included in all programs for pro-moting environmentally and socially responsible aquaculture. Records should be audited and farms inspected, and failure to comply within a specifi ed period would result in loss of the rights given by the government (e.g., use of water, discharge of water, operation of aquaculture facility, etc.) or loss of certifi ed status. Inspections should be random and unannounced so that facilities are less likely to relax their attention to BMPs and standards until time for inspection is near. In instances of noncompliance, certifi ers should clearly state the reason(s) and explain the corrections that must be made to regain compliance.

In the United States, state departments of environmental protection or federal agencies have jurisdiction over nonpoint source pollution control programs and make decisions about the consequences of noncompliance with BMP programs, usually dependent on the severity of the violation. The NPDES program is administered by EPA regional offi ces or delegated to states for enforcement in a manner that provides for remediation before puni-tive action. The GAA certifi cation standards stress continual improvement. Annual review of records is essential for achieving this goal.

Although punitive or corrective action may be necessary in the event of noncompliance, it is important to understand the reasons and motivations for noncompliance. Involving producers in the development of BMPs is critical to assuring compliance. In particular, producers may perceive or experience a conflict between profitability and implementation of BMPs, reinforcing the need to develop cost-benefit analyses for BMPs. Furthermore, during BMP implementation, education and outreach programs need to communicate the full range of benefits, beyond environmental protection, of BMP implementation.

6

Better Management Practices for Freshwater Pond Aquaculture

INTRODUCTION

Most of the fish and crustaceans produced in world aquaculture are cultured in ponds. In the United States, ponds are used to grow channel catfish (*Ictalurus punctatus*) and cray-fish (*Procambarus* spp.), which together comprise almost 80% of the country's annual aquaculture production by weight. Ponds are also used to grow economically important crops of baitfish, hybrid striped bass, freshwater ornamental fish, and penaeid shrimp. Overall, approximately 125,000 ha of ponds are used for aquaculture in the United States.

Aquaculture ponds can be operationally defi ned as confined bodies of standing water that are managed to produce a crop of the target aquatic organism. Ponds are usually envisioned as being constructed entirely of soil, but that is not a necessary part of the definition. Some ponds, for example, may be lined with plastic to reduce seepage.

With respect to possible environmental effects of pond aquaculture, ponds can be func-tionally defi ned as aquatic systems where, by virtue of long hydraulic residence times, suitable water quality for production is primarily controlled by natural physical, chemical, and biological processes. In contrast, suitable water quality is maintained by continuous water exchange in flow-through and net-pen systems and by discrete unit water treatment processes in recirculating systems. This is an important distinction because it suggests that much of the waste loading to ponds is removed from water prior to discharge.

All aquaculture ponds bear a superfi cial similarity, yet potential environmental impacts and "sustainability" may vary considerably. Factors affecting the environmental perfor-mance of pond culture systems include culture intensity, pond hydrological type, and the species cultured.

POND CULTURE INTENSITY

Culture system intensity is an imprecise typology based on a continuum of management inputs and crop yield. The term is often used to classify culture system types; for example, fish production in cages is generally considered more "intensive" than production in ponds. Culture intensity is also used to describe production characteristics within a system type— particularly for pond aquaculture. Extensive pond culture systems have limited water exchange, rely mainly on production of natural foods to support growth of the culture species, and consequently have relatively low crop yields. At the other extreme, intensive pond systems may have higher rates of water exchange, rely on manufactured feeds (often with high levels of animal protein) to promote rapid fish or crustacean growth, and have relatively high crop yields. Semi-intensive systems are intermediate between these inten-sity extremes.

The relationship between culture intensity and sustainability is not direct. Superfi cially, extensive aquaculture might seem more sustainable. However, measures of resource use efficiency in production (such as land, water, fossil fuel energy, and nutrient inputs) are important indexes of environmental sustainability, and efficiency varies among culture systems depending on the particular index examined. For example, production per unit fossil fuel energy input is lower in intensive than in extensive systems, but production per unit land area in extensive systems is much lower than in intensive systems. Although it is difficult to combine the various efficiency measures used to quantify sustainability, intermediate levels of culture intensity are thought to be the most sustainable. Environmental sustainability must be evaluated across a spectrum of criteria and should not be based on a simple set of operational features. Furthermore, sustainability includes economic and social domains, as well as an environmental domain. Extensive pond aquaculture systems generally are not economically sustainable in the United States and all commercial pond aquaculture is based either on semi-intensive culture of "commodity" products (e.g., catfish) or intensive culture of high-value products (e.g., ornamental fish).

POND HYDROLOGY

Pond Hydrological Types

Classifi cation of aquaculture ponds by hydrological type is important because water use, effl uent volume, and, to a lesser degree, effl uent quality are impacted by pond hydrology. Pond hydrology also affects effl uent management practices and the degree to which effl u-ents can be controlled.

Embankment ponds (Fig. 6.1) are the most common type of pond used in aquaculture and are constructed in fl at areas by scraping soil from the pond bottom and using that soil to form embankments around the pond perimeter. Alternatively, the original bottom eleva-tion is maintained for most of the pond area, and the soil used to construct embankments is removed from a trench excavated adjacent to the embankment. Catchment areas are small, consisting only of the pond surface and the inside embankment slopes, so there must be a source of pumped water to fill ponds and maintain water levels during droughts.

Excavated ponds are similar to embankment ponds with respect to general construction and primary water source, but they are usually smaller and the pond bottom is further below the original ground level than in embankment ponds. In some areas with high water tables, excavated ponds may extend below the water table and be partially filled with groundwater inflow. Water may have to be pumped from excavated ponds to empty them.

Watershed ponds (Fig. 6.2) are built in hilly terrain by damming a temporary or per-manent stream. The major source of water is runoff from the drainage basin above the dam, although a source of pumped water is often available to help maintain water levels during droughts. The fourth pond type is a hybrid between embankment and watershed ponds. These ponds may have two or three sides consisting of embankments (actually low dams) across a relatively small, shallow drainage basin. A significant amount of water may be obtained from runoff, but a source of pumped water also must be available because the catchment area above the pond is relatively small. Hybrid watershed-embankment ponds are built in regions with gently rolling topography that is not ideally suited for embankment ponds or watershed ponds.

Embankment and excavated ponds have much less overflow than watershed ponds, with the overflow volume from hybrid ponds being intermediate. Also, the quality of effl uents from watershed ponds may be affected (either positively or negatively) by upstream water quality, which is, in turn, affected by land-use practices in the catchment area.

Water Budgets

Water budgets can be used to identify and describe the magnitude of inflows, outflows, and changes in water level in ponds. The basic water budget can be described as: change in storage (pond depth) = inflows - outflows. For ponds, the budget may include some or all of the following: inflows = precipitation, runoff, stream inflow, groundwater inflow, and regulated inflow; outflows = evaporation and evapotranspiration, seepage, overflow, consumptive use, and regulated discharge.

Water budgets are influenced mainly by climatological variables, particularly rainfall and temperature. Most pond water budgets have been developed to estimate discharge volume because of interest in the possible impacts of effl uents on downstream ecosystems. However, the general hydrological equation can easily be rearranged and solved for esti-mates of water use, rather than overflow. Water budgets have been prepared for excavated or embankment ponds (Boyd 1982; Pote et al. 1988; Teichert-Coddington et al. 1988; Green and Boyd 1995; Nath and Bolte 1998; Braaten and Flaherty 2000) and watershed ponds (Shelton and Boyd 1993; Boyd et al. 2000).

Accounting for water in aquaculture has two significant implications. First, water is a resource that must be used wisely to assure sustainable development. Second, water budgets can be used to estimate effl uent volume and develop options for managing dis-charged water. This is important because water discharged from ponds may impact the water body receiving the effl uent and the extent of that impact is greatly influenced by effl uent volume. These two issues—conservative water use and minimizing effl uent volume—are inti-mately linked. Water that is discharged from a facility is a loss process on one side of the water budget equation. By decreasing the magnitude of that loss, less water will be needed on the other (input) side of the equation. In this section we discuss water budgets for aquaculture ponds, with the goal of pointing out how water can be used more effi ciently and how effl uent volume can be reduced.

Water Sources for Ponds

The main water sources for aquaculture ponds are precipitation, runoff from watersheds, and regulated additions of water from groundwater aquifers or surface water bodies (e.g., streams and reservoirs). Precipitation and runoff are the most important water sources for watershed ponds; precipitation and regulated inflows from groundwater or surface water are the primary water sources for embankment and excavated ponds.

Precipitation

Precipitation falling directly into a pond can be a significant source of water in humid climates. The increase in pond water level from direct precipitation is equal to rainfall. The frequency distribution of precipitation events is approximately log-normally distrib-uted: there are many more precipitation events of low amount than of high amount.

In the lower Mississippi River valley, average annual precipitation is about 125 to 150 cm and ranges from about 6 to 9 cm/month in the fall to about 14 to 16 cm/month in the spring. Long-term average rainfall obscures seasonal variability and rainfall extremes. The range of monthly rainfall among years is great, particularly during winter and early spring. Variability in rainfall is greater in the summer than in the spring. Most tropical areas are characterized by distinct rainy and dry seasons, which have profound impacts on pond water budgets.

Providing storage volume in ponds to capture rainfall (or runoff) will minimize the need for pumped water, particularly during the summer, and reduce effl uent volume released during pond overflow.

Many producers make some effort to capture this free water source. Boyd and Gross (2000) recommend maintaining a water storage capacity equal to the normal maximum daily precipitation.

Watershed runoff

Once the capacity of soil to hold moisture (infi ltration capacity) is exceeded, water will begin to flow across the land surface. Watershed ponds are constructed to capture overland flow from the surrounding watershed, especially if groundwater or surface water avail-ability is not reliable or predictable. The volume of runoff will depend on watershed characteristics, particularly slope, type and extent of vegetative cover, soil type, and ante-cedent moisture. The volume of runoff is also a function of rainfall amount and intensity. Methods for estimating the proportion of rainfall that is converted to runoff are reviewed by Yoo and Boyd (1994). Across the United States, runoff averages 28% of precipitation. In the southeast United States, overland flow is about 21 to 29% of rainfall from November to March, but very little runoff occurs from April to October (Yoo and Boyd 1994). The total annual runoff from various soil areas of Alabama ranges from 46 to 65 cm (Yoo and Boyd 1994).

Approximately 6.5 to 9.8 ha of watershed is recommended to support 10,000 m^3 of water storage in the southeastern United States (USDA-SCS 1982). Many ponds in west Alabama and east Mississippi are constructed as a combination of watershed and embank-ment ponds, and many catfish farms in this area have wells to supplement inflow from runoff. Fluctuations in water level in watershed ponds will exceed those in levee ponds because runoff is a much more important source of water for watershed ponds than levee ponds. During droughts, water levels in ponds supplied only by runoff may decline to dangerously low levels.

Regulated inflows

Regulated inflows of water are provided to fill ponds initially, replace losses from evapo-ration and seepage, and exchange water to manage water quality. The main sources of regulated inflow water are groundwater aquifers and surface water supplies. Water can be added by pumping or gravity flow.

Groundwater

Groundwater has favorable and unfavorable characteristics for use in aquaculture ponds, but on balance it is an excellent source of water for pond aquaculture. Groundwater is usually free of the chemical and biological contaminants that may be present in surface waters. Groundwater is usually not subject to the same fl uctuations in availability as surface waters. However, groundwater is usually cooler than surface waters, has no dis-solved oxygen, and may have elevated concentrations of reduced substances that are toxic to fish.

Groundwater aquifers are underground reservoirs of water that are usually associated with pervious alluvial layers of soil. Favorable attributes of groundwater aquifers that are suitable as a water supply for aquaculture ponds include shallow depth to aquifer surface, high transmissivity and hydraulic conductivity, abundant alkalinity and hardness, separa-tion from potential surface contamination by an impervious layer, and recharge that main-tains aquifer volume. These attributes are characteristic of most aquifers that supply water to ponds in the main catfish growing areas in Mississippi, Arkansas, and Louisiana, as well as the main crayfish growing areas of south Louisiana.

Surface waters

Water can be pumped into ponds from streams and rivers to supplement groundwater and runoff sources. However, surface waters are undesirable from many standpoints. Generally, the seasonal availability of surface waters does not correspond to demand for pond aquaculture. In the southern United States, peak stream discharge occurs during winter to early spring, yet the demand for surface water, particularly as a supplemental water source, is maximum during the summer when stream discharge is near the annual minimum. The seasonal and year-to-year availability of surface waters is extremely variable. In extremely dry years, insufficient surface water may be available to supplement other water sources. In extremely wet years, fl ooding of streams and rivers may damage pond levees and drain-age structures. Overtopping of levees by fl oodwaters may result in escape of cultured stock. Surface waters are also undesirable because they are sources of contamination by wild fish, potential disease pathogens, agricultural pesticides, fertilizers, and sediment. Finally, in the United

States, surface waters are considered Waters of the State and are therefore subject to potential regulatory constraints that may limit withdrawals for pond aquaculture.

Water Losses from Ponds

Water loss from ponds results from evaporation, seepage, and overflow that occurs when inflows exceed pond water storage capacity. The sum of evaporation and seepage repre-sents consumptive water use because this water is not available for downstream uses. Water is also lost when ponds are intentionally drawn down or drained for harvest, pond renovation, or in the case of crayfish culture, to stimulate crayfish burrowing and prepare ponds for planting vegetation. Water that overflows or is intentionally discharged is avail-able for other purposes and thus does not represent a consumptive use.

Evaporation is controlled by climatic forces beyond the control of the culturist. However, seepage can be reduced by good site selection and pond construction practices, overflow can be minimized by maintaining water storage capacity and not managing water quality with water exchange, and losses from draining can be minimized by reducing the fre-quency of pond draining.

Evaporation and evapotranspiration

Pond evaporation is the loss of water from the pond surface to the atmosphere. The rate of evaporation increases as wind speed and the difference in vapor pressure between water and air increases, and decreases as the latent heat of vaporization and water density (both a function of water temperature) increases. Evaporation rates tend to follow annual tem-perature cycles, so evaporation rates are maximum during summer and minimum during winter in the temperate zone.

As a practical matter, pond evaporation is difficult to measure directly because decreases in pond water stage are due to the combined effects of evaporation and seepage, which cannot be easily differentiated. Accordingly, pond evaporation rates are estimated indi-rectly by multiplying the evaporation rate determined with a class A evaporation pan by 0.81 (Boyd 1985a). Measurements from class A evaporation pans (sometimes simply called *pan evaporation*) can be obtained from weather monitoring stations located through-out the world. Average monthly pan evaporation in the southeastern United States ranges from about 2.8 cm/month (0.9 mm/day) in January to 19.7 cm/month (6.6 mm/day) in June (Boyd and Tucker 1998). Over the period 1962 to 1986, maximum daily pond evaporation rates in the lower Mississippi River valley during June ranged from 8.9 to 11.4 mm/day (Pote et al. 1988). Pond evaporation from excavated ponds in Panama was 1.4 to 8.4 mm/ day (Teichert-Coddington et al. 1988). In Thailand, pond evaporation from brackishwater ponds was 3.1 mm/day (Braaten and Flaherty 2000) and from freshwater ponds was 4.5 mm/day (Nath and Bolte 1998).

In some ponds, rice or other vegetation is planted or grows naturally as a forage base for the culture of crayfish. Transpiration by vegetation increases the water loss compared to evaporation by open ponds. In practice it is difficult to separate evaporation from tran-spiration, so the two processes are grouped together. Coeffi cients that depend on the vegetative growth stages of rice are applied to evapotranspiration from a reference crop (grass or alfalfa). When a crop is fi rst planted, nearly all of the evapotranspiration is from evaporation. At full cover, about 90% of evapotranspiration is from transpiration. Estimated evapotranspiration rates for crayfish ponds range from 1.2 mm/day in December to 5.4 mm/ day in August (Parr 2002).

Seepage

Ideally, ponds should be constructed in soil that is impervious to water movement. In practice, however, water moves into or out of all ponds unless they are lined with a water-tight material, such as plastic. Although water may move into ponds from the surrounding soil when local water tables are near the soil surface, the net movement of water through most pond soils is generally outward, and this loss is called *seepage* or *infi ltration*. Seepage is difficult to measure because soils are not uniformly impervious across the pond bottom and evaporative losses are difficult to partition from seepage losses. Generally, seepage is estimated by subtracting pond evaporation from total losses measured by decrease in water stage.

Embankment ponds in the Mississippi River valley are constructed in montmorillonite clay soils that are dense and swell when wet, becoming nearly impervious to water flow. Seepage from rice fields and crayfish ponds constructed in these soils is 0.38 mm/day (Pringle and Pennington 1991). Limited measurements from embankment ponds con-structed in the lower Mississippi River valley indicate that seepage is <1.3 mm/day (Pote et al. 1988). Seepage from ponds constructed in the Black Belt soil edaphic region of east-central Mississippi and west-central Alabama is 1.5 mm/day (Parsons 1949). Seepage was calculated for excavated ponds in Thailand (4.4 mm/day) and Honduras (5.8 mm/day) (Nath and Bolte 1998) and for brackishwater shrimp ponds on the central plain of Thailand (5.2 mm/day) (Braaten and Flaherty 2000). Boyd (1986) classifi ed seepage rates as low (0 to 5 mm/day), moderate (6 to 10 mm/day), high (11 to 15 mm/day), and extreme (>15 mm/day). About 87% of the water lost from ponds at a research station in Panama was from seepage, which ranged from 19 to 58 mm/day (Teichert-Coddington et al. 1988).

Excessive seepage is undesirable because it increases the need for pumped water. Also, excessive seepage can pose a threat to the pristine quality of underground water supplies because uncharged substances or anions such as nitrate can percolate through soils with seepage water. Seepage also has a profound effect on the amount of water lost as overflow. In ponds with low seepage rates, a proportionally greater volume of water is discharged as overflow. In ponds with high seepage rates, water is lost to infi ltration and overflow rates are low.

Overflow

Overflow occurs when pond inflows exceed pond storage capacity. Overflow is intentional when water quality in ponds is managed with water exchange. Overflow is unintentional when inflows from rainfall and runoff exceed pond storage capacity. In embankment ponds, the volume of unintentional overflow depends on the capacity of the pond to store pre-cipitation. In a year with average precipitation in northwest Mississippi, overflow from ponds managed with 15 cm of storage capacity is about 33 cm (Tucker et al. 1996). Estimated overflow ranges from 20 to 38 cm for ponds managed with 15 to 46 cm of storage capacity (Tucker et al. 1996; Hargreaves et al. 2001). The volume of unintentional overflow from watershed ponds is much greater (235 to 380 cm) because of the much greater contribution of runoff to the water budget of watershed ponds (Boyd et al. 2000). Most unintentional pond overflow occurs from fall to mid-spring, the cooler, wetter part of the year in the southeast United States, when precipitation exceeds losses from evapo-ration and seepage. In years with normal rainfall, 72% of overflow from watershed ponds occurs between December and April (Shelton and Boyd 1993). Very little overflow occurs during the dry summer months when evaporation often exceeds precipitation. Pond over-flow volume can be reduced by maintaining some storage capacity for rainfall and runoff.

Water exchange is no longer used to manage water quality in most ponds for finfish in the United States. However, water exchange is used to manage dissolved oxygen in the shallow ponds used for crayfish culture. From 1 to 9 water exchanges (or 50 to 450 cm) may be used during one crop cycle.

Pond draining

Ponds are drained to facilitate harvest, adjust fish inventory, remove accumulated sediment or lime pond soil, and repair embankments. The volume of water discharged when ponds are drained is equivalent to that added when ponds are filled. Thus, from the perspective of water use, there is no net loss of water as a result of filling and draining ponds. However, from the perspective of effl uents, substantial changes in water quality occur during culture, with substances either accumulating during the culture period, reaching an equilibrium in response to feeding, or refl ecting conditions in the short period prior to discharge.

Long term, the volume discharged from pond draining is a function of draining fre-quency, which varies by species, phase of production, and hydrological pond type. Most foodfish production ponds for channel catfish are drained every 3 to 15 years. Shallow (50 cm) crayfish ponds are drained annually to stimulate crayfish burrowing, an essential part of the production cycle. Ponds for the culture of phase I hybrid striped bass are drained after 45 to 60 days.

With channel catfish, fi ngerling ponds are drained annually, broodfish ponds are drained every 1 to 5 years, and foodfish ponds are drained every 3 to 15 years (USDA-APHIS 2003). Although fi ngerling ponds are about 13% of the total area in catfish production, they contribute 47% of the total volume discharged from ponds when ponds are drained. Because fi ngerling ponds represent an inordinate share of the water discharged from catfish ponds, considerable scope for water conservation and reducing the volume of effl uent in channel catfish aquaculture may be derived by improving water management in fi ngerling ponds.

Embankment or excavated ponds represent over 90% of the ponds used for catfish culture (USDA-APHIS 1997). Final fish harvest occurs only when ponds need to be drained for renovation or adjustments in fish inventory are deemed necessary. In contrast, watershed ponds may be partially or completely drawn down to facilitate fish harvest. Partial drawdown occurs about every 6 years and complete drawdown occurs about every 15 years. In 15 years, there will be two drawdowns of 50% of the pond volume and one drawdown of 100% of the pond volume (total = 2 pond volumes).

For embankment or excavated ponds drained annually or more frequently, the volume of water discharged from pond draining can exceed that discharged by unintentional over-flow. For ponds with a long interval between draining, such as foodfish catfish ponds, most of the effl uent is discharged from unintentional overflow. Thus, one obvious way to minimize effl uent volume is to extend the interval between pond drainings.

MAJOR SPECIES CULTURED IN PONDS

Ponds are used to grow a wide range of species, including channel catfish, baitfish, ornamental fish, hybrid striped bass, bluegill (*Lepomis macrochirus*), largemouth bass (*Micropterus salmoides*), walleye (*Sander vitreus*), trout, crayfish, penaeid shrimp, fresh-water prawns (*Macrobrachium rosenbergii*), and many other freshwater and saltwater aquatic animals. Potential environmental impacts can vary widely depending on culture species, husbandry practices, and pond hydrology. In fact, culture conditions and potential impacts of peneaid shrimp culture differ to such an extent that shrimp ponds are considered in a separate chapter (Chapter 7).

The diversity of species and culture methods employed in pond aquaculture make the development of generic environmental management practices particularly challenging. A reasonable practice for one species may be totally inappropriate for another, and some practices are more appropriate for a particular facility than other practices. Furthermore, practices may be combined in unique ways to achieve the goal of pollutant reduction. The following sections will present brief overviews of the most important species cultured in ponds, focusing on water use, discharge frequency, and effl uent volume.

Channel Catfish

Other than perhaps rainbow trout, more is known of the biology and culture of channel catfish than any other aquaculture species grown in the United States. Moreover, environ-mental management of catfish farming has been more thoroughly studied than that of any other species. General aspects of catfish farming are thoroughly reviewed in Tucker and Robinson (1990) and Tucker and Hargreaves (2004). Catfish pond effl uents are character-ized by Boyd (1978), Tucker and Lloyd (1985), Tucker and van der Ploeg (1993), Schwartz and Boyd (1994), Shireman and Cichra (1994), Tucker et al. (1996), Boyd and Tanner (1998), Boyd and Gross (1999), Boyd et al. (2000), Tucker et al. (2002), and Hargreaves et al. (2005a, b). General pond ecology and identifi cation and management of environmental impacts of catfish farming are described by Schwartz and Boyd (1996), Tucker (1996), Boyd and Tucker (1998), Boyd et al. (2000), Tucker et al. (2002), Hargreaves and Tucker (2003), and Tucker and Hargreaves (2003).

Channel catfish are grown commercially in at least 35 states, although more than 95% of all United States catfish production is derived from farms in Alabama, Arkansas, and Mississippi. Within those states, production occurs in two well-defi ned geographic regions:

1) the Mississippi River fl oodplain in northwest Mississippi and southeast ,
2) the Blackland Prairie region of west-central Alabama and east-central Mississippi. These two catfish-growing regions

are separated by about 250 km and have similar cli-mates, but pond hydrology and water use differ considerably between the two regions.

About 50,000 ha of ponds are used for catfish aquaculture in the Mississippi River Alluvial Valley. Nearly all the catfish fi ngerlings used in catfish farming are grown in this region. The land is fl at and ponds are constructed on clay soils that have low infi ltration rates (<0.1 cm/day). Embankment ponds are universally used for aquaculture in the region, and they are built by removing soil from the area that will be the pond bottom and using that soil to form embankments around the pond perimeter. Ponds range in size from 2 to 6 ha and are 1 to 2 m deep. Water for filling ponds is pumped from the Mississippi River Alluvial Aquifer. After ponds are initially filled with groundwater, water levels are main-tained by inputs of precipitation and pumped water. Abundant rainfall in the region sig-nifi cantly reduces the requirement for pumped water after ponds are filled (Pote et al. 1988).

An additional 25,000 ha of catfish ponds are located in west-central Alabama and east-central Mississippi, mostly in the western portion of the Blackland Prairie soil edaphic region. The land is nearly fl at to moderately hilly. High-yielding aquifers in the Blackland Prairie are considerably deeper than the Alluvial Aquifer along the Mississippi River, and this is refl ected in water-use patterns and pond types in the prairie region. Whereas all catfish ponds in the Mississippi River fl oodplain are embankment ponds filled with ground-water, about 75% of the commercial catfish ponds in the Blackland Prairie are watershed ponds that use mainly rainfall and storm runoff for filling and maintaining water levels. Ponds are designed to maintain a minimum water level with average rainfall. The remain-ing 25% are embankment ponds similar to those used in the Mississippi River fl oodplain. These ponds are filled with pumped water from wells or surface water from adjacent creeks. Some ponds in the Blackland Prairie are hybrid watershed-embankment ponds built with two or three embankments that confine water against a slope in a broad water catchment area. After initial filling, pond water levels are maintained with precipitation, runoff, and pumped water, depending on availability.

Catfish aquaculture is usually practiced in four discrete phases: 1) broodfish are held in ponds at relatively low standing crops and allowed to randomly mate each spring; 2) fertilized eggs are taken from the broodfish pond to a hatchery where they hatch under controlled conditions; 3) fry are transferred from the hatchery to a nursery pond where they are fed a manufactured feed for about 6 months; 4) fi ngerlings are moved from the nursery pond to foodfish production ponds where they are fed a manufactured feed until they reach 0.7 to 1 kg. In the southeastern United States, 18 to 30 months (two or three growing seasons) are required to produce a food-size channel catfish from an egg.

Water use varies considerably with each phase of catfish farming. Broodfish ponds (which represent about 2% of the total pond area devoted to catfish production) are typi-cally drained every 2 to 5 years for inspection and replacement of broodfish. Hatcheries operate only from late April through early July, corresponding to the natural catfish breed-ing cycle. Catfish hatchery water use and effl uent characteristics are described by Tucker (2005). Fingerling nursery ponds (about 13% of total catfish pond area) are usually drained every year after fish have been harvested and moved to foodfish ponds.

Foodfish-production ponds constitute about 85% of the total pond area used to raise catfish. Most foodfish ponds are used for several years before being drained. Market-sized fish are selectively harvested using large-mesh seines and fi ngerlings are added to replace the harvested fish (Fig. 6.3). The process of selective harvest and "understocking" can continue for many years without draining the pond. Some producers use a variation of this cropping system in which as many fish as possible are harvested after a growing season without draining the pond. The pond is subsequently restocked with fi ngerlings and grow-out continues. Foodfish grow-out ponds are usually drained only when pond levees need to be renovated or when there is need to adjust fish inventory by completely harvesting the crop. Commercial foodfish ponds remain in production for 3 to 15 years, or more, without being drained; the average time between pond drainings is more than 9 years (USDA-APHIS 2003). Intentional water exchange (fl ushing) is not routinely practiced in catfish ponds.

Catfish standing crops and feeding rates are relatively high in foodfish ponds. Annual production ranges from 6,000 to 7,000 kg/ha (5,000 to 6,000 pounds/acre), and feed inputs during the summer are routinely in the range of 100 to 150 kg/ha per day. That level of

production is possible in ponds with limited mechanical aeration and no water exchange only because natural physical, chemical, and biological processes within the pond assim-ilate or remove wastes that are generated as a result of nutrient input from feeds. Although catfish production has steadily increased over the last 40 years, it appears that the loading limits of the production system as currently confi gured and managed have been reached. Factors affecting the loading limits of catfish ponds are detailed by Hargreaves and Tucker (2003).

Red Swamp Crayfish

Culture of the red swamp crayfish (*Procambarus clarkii*) and white river crayfish (*P. zonangulus*)—collectively referred to as *crawfish* in the southern United States—ranks second only to channel catfish in terms of area devoted to aquaculture ponds in the United States. In 2005, there were about 47,000 ha of crayfish ponds in Louisiana (Lutz and Romaire 2005) and since the mid-1990s the production area has ranged from 45,000 to 52,000 ha (Avery and Lorio 1999). Crayfish farms in Louisiana produce about 90% of the crayfish in the United States. Less than 3,000 ha of crayfish ponds are located outside of Louisiana (McClain and Romaire 2004), with most of that area in Texas, Florida, and South Carolina. The main crayfish growing areas of the coastal plain and river alluvial valleys of southwest and south-central Louisiana overlap with the main rice-growing areas. The basic aspects of crayfish biology and culture are reviewed by Huner and Barr (1991), Avery and Lorio (1999), and Eversole and McClain (2000). Crayfish pond effl uents were characterized by Orellana (1992) and effl uent mass discharge was modeled and measured by Parr (2002). Specifi c environmental management practices for crayfish farming have been compiled by Lutz et al. (2003).

Production of crayfish in Louisiana is unique to pond culture in the United States because ponds are operated at low intensity, with yield strongly related to the development of a detrital food web based on natural plant forage and factors affecting natural crayfish recruitment. Unlike most pond culture of finfish in the United States, manufactured feeds are not applied to crayfish ponds. Hatcheries are not necessary for production of juveniles. Crayfish are not harvested by seine, but are captured gradually over the course of the production season in baited traps. Average production is low, around 600 kg/ha, with annual yields ranging widely from 100 to 1500 kg/ha because of variation in type and biomass of forage, the availability of and ability to manage water, and factors affecting natural recruitment of young crayfish.

Water is managed to mimic the hydrological cycles to which natural populations of crayfish are exposed. Water use and management is dependent on the type of production system used for crayfish culture. With permanent ponds, crayfish are cultured in ponds constructed specifically and solely for crayfish culture. Typical ponds are 4 to 8 ha in surface area with a shallow water depth that ranges from 20 to 60 cm (Fig. 6.4). A forage crop such as rice is planted in summer, or naturally occurring moist-soil plants are allowed to grow. The pond is fl ooded in October and crayfish are harvested from late fall to late spring. Permanent ponds are drained in late spring or early summer and the cycle repeated. With rotational ponds, agronomic crops such as rice and soybeans are planted in rotation with crayfish. Water management in rotational ponds represents a compromise between the demands of the crops in the rotation. For example, in the rice-crayfish-rice rotation, rice is planted in spring and the pond fl ooded to a depth of 20 to 25 cm.

The pond is drained in late summer to allow mechanical rice harvest. Forage produced from the remaining rice stubble—the ratoon crop—is used as the forage base for crayfish culture. Ponds are fl ooded in October and crayfish are harvested from late fall to early spring, when ponds are drained in preparation for rice planting. Yields from rotational ponds tend to be lower than those from permanent ponds because forage biomass is usually lower and the duration of crayfish harvest is shorter. In both permanent and rotational ponds, ponds are drained annually over 10 to 14 days, usually in the spring. Ponds are deliberately drained slowly to promote the construction of crayfish burrows that are sufficiently deep to avoid drought during summer and enhance survival of burrowed crayfish.

Much more water is used for water exchange than for initial pond filling because water exchange is commonly used to manage water quality in crayfish ponds. When ponds are fi rst fl ooded in October, decomposition of vegetation can lead to dissolved oxygen deple-tion, a condition that is managed as necessary by exchanging pond water with pumped surface or groundwater. Depending on the density of vegetation and the severity and dura-tion of low dissolved oxygen concentration, as many as 7 to 9 pond volumes may be exchanged over a production season (Avery and Lorio 1999). Alternatively, crayfish farmers can manage dissolved oxygen by recirculation and aeration of pond water with paddlewheel aerators in ponds confi gured with internal baffl e levees and access to electri-cal power. The cost of electricity or fuel for pumping is a major expense in crayfish farming, so there is a strong economic incentive to reduce the use of pumped water to manage water quality.

Baitfish

Baitfish farming is a significant aquaculture industry in the United States, with perhaps 15,000 ha of ponds devoted to baitfish production. Although many species are raised for bait and culture conditions vary widely, three species (golden shiners, fathead minnows, and goldfish) constitute over 90% of overall sales and culture practices for those three species are somewhat standardized. Golden shiners (*Notemigonus crysoleucus*) and fathead minnows (*Pimephales promelas*) are used as fishing bait, while goldfish (*Carassius auratus*) and larger fathead minnows are primarily sold as food for carnivorous ornamen-tal fish. Reviews of baitfish farming are provided by Stone (2000) and Stone et al. (2005). Baitfish pond effl uents are characterized by Bodary et al. (2004) and Frimpong et al. (2004).

Baitfish may be captured from the wild or cultured in ponds. About half of all baitfish sold are wild-caught and in some regions of the upper northeast and midwestern United States, the percentage of wild-caught bait is much higher. Harvesting baitfish from the wild may impact natural fish communities by overharvesting or by depleting forage for game species, although the greatest concern is the widespread transportation of baitfish and the potential for introducing species into ecosystems where they are not indigenous. Farm-raised fish are a substitute for wild fish and therefore reduce the possible impacts of obtaining bait from the wild. Farming provides a reliable source of a few species that are already widely distributed and generally innocuous.

Most farm-raised baitfish are grown in managed ponds in the southeastern United States. In 1998 (the latest fi gures available) there were 275 baitfish farms in the United States (USDA-NASS 2000) with total annual sales exceeding $40 million (Collins and Stone 1999). Arkansas is the largest producer of farm-raised baitfish, accounting for more than 60% of total sales.

Most baitfish are grown in levee-style earthen ponds that are hydrologically similar to catfish ponds in the Mississippi River fl oodplain. Ponds for golden shiners and fathead minnows are 2 to 10 ha and ponds for feeder goldfish are usually less than 1 ha. Water is supplied from wells because surface waters contain wild fish and disease organisms. Ponds often are filled only to a depth of 25 to 50 cm for best egg hatch or fry survival. Water from adjacent ponds is then added to seed the new pond with plankton. After ponds are filled, water is added only to replace evaporation and seepage. Intentional water exchange is rare because dilution will reduce the abundance of natural foods, which are an important source of nourishment for baitfish, even if manufactured feeds are being used. Serious groundwater supply issues in the major baitfish-producing region of Arkansas during the 1980s and 1990s prompted most baitfish farmers in that region to adopt water-conservation practices that include formal pond water-level management to capture rainfall and reusing pond water discharge whenever possible.

Fathead minnows and golden shiners are usually harvested multiple times over the growing season to supply bait markets. Some ponds are drained each year to make room for the next year's crop. Some ponds are not drained because fish are needed to meet spring and early summer demand for bait. These ponds are stocked by transfer of fish from overcrowded nursery ponds. Feeder fish production is different because ponds must be totally harvested over a short period before fish become too big for the feeder market.

Standing crops of baitfish in ponds are low compared to catfish aquaculture. Maximum standing crops of golden shiners and fathead minnows may be 400 kg/ha, or less than one-tenth of the biomass in catfish ponds. Maximum standing crops of goldfish are higher than for other baitfish and may be as high as 800 kg/ha. Waste generation in baitfish ponds is therefore relatively low compared with other pond aquaculture.

Hybrid Striped Bass

Interest in culturing striped bass and its hybrids for food was stimulated by dramatic declines in the Atlantic coast wild fishery for striped bass (*Morone saxatilis*) during the 1980s. Striped bass are an esteemed foodfish along the Atlantic seaboard and striped bass aquaculture grew steadily from the mid-1980s to fill established commercial markets. Popularity of the fish has also increased in response to the greater availability provided through aquaculture. In 2005, 5,400 tonnes of hybrid striped bass were produced in the United States, with about 3,200 tonnes produced in approximately 600 ha of ponds, primarily in Texas, Mississippi, and North Carolina. Most of the remainder was produced in intensive flow-through tanks or recirculated aquaculture systems in the northeast, midwest, and west. Although there may be more than 70 facilities growing hybrid striped bass in the United States, production is highly skewed toward only a handful of large operations. In 2000, 60% of the total annual production was grown by only three operations.

Striped bass are not a particularly good fish for pond aquaculture, and all commercial production uses hybrids made by crossing striped bass with white bass (*Morone chrysops*). The original cross, or palmetto bass, is made by crossing female striped bass with male white bass. The reciprocal cross, or sunshine bass, is most commonly used in aquaculture and is made by crossing female white bass and male striped bass. Both hybrids grow faster and tolerate aquaculture conditions better than either parent. Compilations of information on the biology and culture of hybrid striped bass are available in Harrell et al. (1990), Kerby (1993), Harrell (1997), Kohler (2000), and Woods (2005).

Pond culture of hybrid striped bass consists of several distinct phases, beginning in early spring when white and striped bass are crossed in a hatchery. After 2 to 4 days in the hatchery, fry are transferred to fertilized rearing ponds where they remain for 30 to

45 days. Fish are fed a fi nely ground manufactured feed during the last half of this period to reduce cannibalism and to wean fish (called *phase I fi ngerlings*) onto manufactured feed. Fingerlings are then harvested as 1-g fi ngerlings, graded to uniform size, and restocked in phase II ponds for further growth. Waiting longer to harvest and grade phase I fish results in poor survival due to cannibalism. Relatively few operations produce feed-trained phase I fi ngerlings and those that do supply most of the industry.

Most farmers buy feed-trained, size-graded fi ngerlings and grow the fish to market size in either one or two steps. When grown in one step, farmers stock fi ngerlings at 1,200 to 2,000 fish/ha and fish are fed and grown to market size (0.5 to 1 kg) over a 15- to 18-month period. In the second, and more common, production option, farmers allocate about 25% of the farm's pond area to an intermediate step where phase I fi ngerlings are stocked at high densities (6,000 to 12,000 fish/ha), fed, and grown until fish are 1 year old and weigh 100 to 150 g/fish. Fish from phase II culture are then harvested, graded to uniform size, and restocked into ponds at 1,200 to 2,000 fish/ha for growth to market size (phase III culture).

Embankment-type ponds initially filled with well water are used for all posthatchery culture phases. Ponds used in phase I culture are 1 to 2 ha and 1 to 2 m deep; ponds used for phase II culture and fi nal grow-out are usually larger (2 to 5 ha). A major difference between ponds used for catfish culture and those used for hybrid bass culture is the increased reliance on water exchange to maintain good water quality in ponds used for phase II and III culture (or for fi nal grow-out in the two-step process).

Water exchange is particularly common toward the end of fi nal grow-out when feeding rates are highest. Pond managers often use a predetermined water-quality criterion (such as total ammonia-nitrogen levels or water temperatures) to initiate fl ushing, but water may be exchanged whenever the pond manager feels that fish health is threatened. Water use varies greatly among ponds and among facilities. Based on anecdotal reports, ponds may be fl ushed several times during grow-out, with overall water use totaling one or two pond-volumes per year.

Ponds are always drained between crops of phase I and phase II fish, and most manag-ers drain ponds between crops of foodfish. However, foodfish ponds on at least one large hybrid striped bass farm are not drained between crops as a water conservation and time-saving measure. On that farm, ponds are used for 4 to 8 years without draining and fish remaining in ponds between crops are killed using a piscicidal chemical.

Ornamental Fish

The trade in ornamental fish for use as pets in home aquaria is an important global indus-try. The worldwide value of the aquarium hobby exceeds $5 billion annually, and the United States is the largest importer of ornamental fish and is second only to Singapore as the largest supplier. Florida is the center for ornamental fish supply in the United States, providing more than 90% of the aquarium fish sold in the United States. Most of the fish sold from Florida are imported from overseas and redistributed. However, a significant aquaculture industry exists in that state for certain species and the 2003 farm-gate value exceeded $47 million. In addition to "tropical" fish cultured mainly in Florida, ornamen-tal goldfish and koi for display in water gardens and landscape pools are grown on small facilities throughout the United States.

Hundreds of ornamental fish species are grown in the United States and it is impossible to summarize the wide variety of production techniques. Some of the commonly cultured groups are in the families Cyprinidae (barbs, danios, goldfish, koi), Characidae (tetras), Callichthyidae (armored catfishes), Loricariidae (plecostomus), Poeciliidae (livebearers), Cichlidae (cichlids), and Belontiidae or Osphronemidae (gouramis) (Hill and Yanong 2006). Information on ornamental fish aquaculture has been summarized by Chapman (2000), Watson and Shireman (2002), and Hill and Yanong (2006). Many of the produc-tion practices used on ornamental fish farms were developed by trial and error and are closely guarded business secrets.

Depending on the species raised, indoor facilities may be used for breeding and larvi-culture; earthen ponds are commonly used for grow-out of juveniles to market size. About 600 ha of ponds are used to produce ornamental fish in Florida. However, ever-increasing land values and dwindling water resources in Florida have lead to increased interest in using intensive, water-recirculation technology for all phases of production. Intensive indoor culture technology also reduces the risk of fish loss to predation and sudden decreases in water temperatures during winter months.

Ornamental fish ponds are much smaller than ponds used for other species. Typical ponds are 25 m long by 6 m wide and 2 m deep (Fig. 6.5). Ponds are usually excavated in sandy soil or, in south Florida, coral bedrock. Ponds are often excavated below the water table and must be pumped dry between crops. Most ponds are also supplied with well water.

their nutrition directly from the feed or indirectly through the activity of the feed as a source of fertilizer. Most ponds are aerated continuously with small surface aerators and are drained (or pumped dry) between crops. Relative to other pond-raised fishes, orna-mental fish grow rapidly to "market size." Ponds are typically filled and drained three times a year to initiate and end individual grow-out cycles.

SITE SELECTION

There is perhaps no greater predictor of the potential success and environmental impact of an aquaculture operation than site selection. Site selection criteria can be divided into two broad categories. First are those factors that constrain construction and operation of ponds in particular locations. Second are those factors that make sites more or less suitable for pond aquaculture. These include the various physical and environmental factors that influence the suitability of a particular site.

More broadly, site selection can include factors such as proximity to hatcheries, fi nger-ling suppliers, feedmills, fish processing plants, and suppliers of other farm inputs; access to transportation, electrical, and communications infrastructure; and access to skilled and unskilled labor in support of fish production and processing. Despite the inclusion of such a wide range of criteria in site selection, this discussion will focus on the physical and edaphic factors of site selection that have a strong effect on the environmental performance of pond aquaculture. These include the availability of suitable terrain and soils in which to construct ponds and water of appropriate quality and quantity to fill ponds and maintain water levels.

Most of the siting criteria important in reducing environmental impacts of pond aqua-culture are also critical in assuring profitability of the farm. For example, selection of sites with proper soils and protection against fl oods will reduce long-term expenses associated with water use and replacement of escaped stock, while also protecting the environment by reducing water use, reducing infi ltration of pond water into groundwater supplies, and preventing escaped animals from entering nearby water bodies. As such, careful site evaluation should be made prior to construction regardless of the legal requirements. Identifi cation of potential problems before construction and either addressing those prob-lems or locating an alternative site is much less expensive and environmentally benign than solving problems by implementing mitigation measures after construction or, in the extreme, having to close the farm and abandon the site.

Until recently, the tools available for site selection consisted of aerial photographs and soil, topographic, and fl ood maps. Coupled with site-specific investigations and inter-views with local residents, these site selection tools were basic. More recently, site selec-tion has become more sophisticated with the advent of geographic information systems (GIS) that can synthesize and weigh relevant factors and site constraints in the form of overlay maps or coverages for each criteria. Areas of each coverage can be weighted according to subjective criteria of relative importance and the fi nal site selection map coded to indicate sites according to relative suitability. Site selection using GIS can be done at a coarse scale and low resolution (1 to 10 km) or at a fi ne scale and high resolu-tion (10 to 100 m). Examples of site selection using GIS are summarized by Nath et al. (2000).

From the standpoint of environmental impacts, particularly from pond effl uents, infor-mation about soil and water resources is most important. Other criteria may have a greater effect on farm profitability (proximity to input sources and markets, for example), but soil and water resource characteristics have a strong effect on the environmental performance of pond aquaculture. Better management practices for site selection described here are based on rating criteria developed for soil and water resources by Hajek and Boyd (1994). Depending on the value of each criteria, limitations are categorized as slight, moderate, or severe. For purposes of the practices discussed below, site selection criteria for soil and water with only slight limitation are considered characteristic of "best" sites. The United States Department of Agriculture (USDA) Natural Resources Conservation Service has published a set of conservation practice standards, useful in site selection and pond construction, that are available on the Internet (www.nrcs.usda.gov/technical/standards/ nhcp.html).

Better Management Practices for Site Selection

Do not site ponds in wetlands or protected areas

Although the availability of water and the fl at topography of wetlands makes them attrac-tive sites for pond aquaculture, wetland soils often contain high concentrations of organic matter (>10%) that are not suitable for the construction of stable embankments. Organic soils are difficult to work with machinery, resulting in relatively high pond construction costs. Organic soils are elastic, which are difficult to compact properly, suggesting that water loss from seepage may be severe. The elasticity of organic soils also means that they have low load-bearing capacity, so embankments built with organic soils will have difficulty supporting vehicles and machinery. Some wetlands have acid-sulfate soils, which can require very high application rates of lime before they are suitable for pond aquaculture.

Riparian wetlands are susceptible to seasonal fl ooding. In the United States, construct-ing ponds in wetlands requires a permit from the United States Army Corps of Engineers. Siting aquaculture ponds in wetlands often results in increased costs of production and long-term management problems. Furthermore, the long-term intrinsic ecological value of natural wetlands (Costanza et al. 1997) may exceed that derived from conversion to aqua-culture ponds. However, natural or constructed wetlands can be used to treat the effl uent from aquaculture ponds before release to receiving waters.

Do not site ponds in areas prone to regular fl ooding

The best sites for the construction of aquaculture ponds are those with slopes less than 2% and often these areas correspond to riverine fl oodplains or valley fl oors that are subject to periodic inundation in the absence of fl ood-protection levees. Floods that overtop aquaculture pond levees can result in loss of cultured animals, contamination of ponds with feral aquatic animals, mixing of potentially poor quality fl oodwaters with pond waters, and damage to pond embankments and other farm infrastructure. Aside from the possible escape of nonnative animals, the major impact of fl oods will usually be catastrophic for the farmer but benign to the environment since pond waters will be greatly diluted by fl oodwaters.

In the United States, the local USDA Natural Resources Conservation Service provides information on historic fl ood levels and recurrence interval, guidance on avoiding sites in fl ood-prone areas, and suggestions for siting ponds to minimize fl ooding of adjacent lands caused by pond construction. Best sites are those where the frequency of fl ooding is 0 to 5 times in 100 years (Hajek and Boyd 1994). In areas with no fl ood-protection levees, the top of perimeter pond embankments should be 50 cm above historic fl ood levels.

Do not site ponds near urban areas, near industrial pollution sources, or in sites with contaminated soils

Aquaculture producers with ponds located in peri-urban areas have a distinct advantage of proximity to large markets of urban consumers and sources of farm inputs. However, the risks associated with contamination are increased relative to ponds sited in rural areas or where zoned for agricultural uses. Effl uents and runoff from urban wastewater treatment plants and industrial facilities can contaminate surface waters used for aquaculture ponds. Seepage from improperly constructed hazardous waste holding ponds at industrial facili-ties can contaminate groundwater resources. Ponds should not be sited near utility right-of-ways because these areas are often sprayed with herbicide to control vegetation. Aquaculture ponds should be sited at least 3 km from urban areas or potential pollution sources.

Soils previously exposed to certain agricultural, industrial, or urban activities may be contaminated with harmful chemicals that may affect the health of the cultured species or accumulate in tissues that will affect product quality or safety. Contaminants may nega-tively affect natural waste removal processes within ponds or be discharged with effl uents. Soils that are being considered for construction of aquaculture ponds should be tested for residual contamination if there is a history of pesticide use in the area. Often, contaminants will accumulate in topsoil, which should be excavated and removed from the site in preparation for pond construction.

Do not site ponds where regulatory or other restrictions apply

Several state or federal agencies may have jurisdiction over land use and pond construc-tion. Contact the local offi ce of the USDA Natural Resources Conservation Service for guidance. Additional site review may be required by the United States Army Corps of Engineers, which administers and enforces provisions of Section 404 of the Clean Water Act. Section 404 regulates, among other activities, conversion of wetlands to farming, including aquaculture ponds. In some areas, land-use zoning may preclude agriculture in general or aquaculture in particular. Ponds cannot be constructed in national, state, or local parks, forests, conservation areas, or monuments.

Select sites with suitable topography for pond aquaculture

Site topography should allow construction of ponds and any required treatment facilities. Typically, ponds are constructed in low-gradient landscapes with average slopes <5%, and the best sites are those where the slope is <2% (Hajek and Boyd 1994). Costs for embank-ment pond construction on low-gradient slopes are relatively low because less soil must be moved to construct levees than on higher-gradient slopes. For watershed ponds, good sites have a topography of gently sloping ridges with a broad valley behind. Best sites are those where the greatest volume of water can be impounded for the least volume of earth fill for the embankment.

Select sites with suitable hydrology for pond aquaculture

It goes without saying that sufficient water must be available to support pond aquaculture. Water requirements depend on the culture methods appropriate for a particular species or life stage, soils and pond construction techniques, and climatic conditions, particularly precipitation and evaporation. Sites with losses from evaporation and seepage that cannot be replaced by available water resources should be avoided. Sites should be selected that make good use of available rainfall as a water source. Best sites have a climate where the sum of losses from evaporation and seepage minus water added from precipitation is less than 25 cm/year (Hajek and Boyd 1994). Furthermore, sites where rainfall is highly sea-sonal should be avoided unless surface waters are available. Best sites are those where the interval between appreciable rainfall is less than 5 days. Monthly and annual averages, minima, and maxima precipitation and evaporation should be determined for each site to estimate water losses and gains. Absolute values as well as year-to-year variation should be assessed.

Water should be available in sufficient volume to support pond aquaculture. At minimum, sufficient water should be available to fill a pond within 10 days, or 250 L/minute per ha (25 gallons/minute per acre). Water can be obtained from surface waters by gravity or pumping, or pumped from groundwater aquifers. Surface waters are less desirable than groundwaters because availability tends to be seasonal, or the water can be contaminated with pollutants or contain wild fish or organisms that are pathogenic to cultured fish. Surface waters should be fi ltered or screened when used as a water supply for aquaculture ponds. The water quality of surface waters can be widely variable, especially during fl oods. Groundwaters have few of these limitations, although groundwater quality is usually not suitable for direct use in aquaculture ponds without aeration. The temperature of ground-waters is usually less variable than that of surface waters, and it is cooler than surface waters in summer and warmer than surface waters in winter. The depth to the aquifer surface, potential well yields, and water quality should be determined if groundwater is used as a water source. This information can be obtained from local maps, well drillers, or the United States Geological Survey.

Select sites with consideration of the downstream impacts of pond construction

The diversion of large streams to supply water to aquaculture ponds should be avoided. Construction of facilities and access roads should not alter natural water flows needed to maintain surrounding habitats and downstream environments such as wetlands and ground-water aquifers. The potential disruption caused by pond construction and operation to downstream multiple-use resources should be considered. Pond construction should not have a negative impact on downstream cultural or historical resources.

Construction of aquaculture ponds can mitigate or exacerbate fl ooding by alterations to the natural hydrological regime, particularly watershed runoff, depending on the size (i.e., storage volume) and position of the ponds in the watershed and physical alterations to natural drainage patterns from pond construction. Although ponds can be constructed in rolling or hilly terrain, embankments will be necessarily large and, unless spillways are constructed properly, are at increased risk for failure during storms.

Watershed ponds should not be located where embankment failure could result in loss of life or damage to residences, industrial buildings, main highways, or railroads, or inter-ruption of public utilities. Design and construction of embankments should be supervised by a qualifi ed engineer or reputable pond construction company. Embankments should be constructed in compliance with local or state dam safety regulations.

Select sites where water supplies are free from contamination

Water supplies for pond aquaculture should be free of substances that may cause poor water quality in ponds and pond effl uents, cause excessive stress or disease in cultured animals, or accumulate in tissues and affect product quality or food safety. Local aerial pesticide applicators should be notifi ed of the location of aquaculture ponds, and measures to prevent contamination of ponds with pesticide drift should be discussed.

The quality of runoff entering watershed ponds will depend on land uses in the drainage area. Drainage basins above watershed ponds should be protected against erosion by maintaining vegetative cover and land terracing if slopes are steep. Drainage areas that are in pasture or forest are best. Alternatively, a 10- to 15-m wide vegetated buffer strip to intercept sediment in runoff can be maintained around pond margins. Additional infor-mation can be obtained in USDA Natural Resources Conservation Service Conservation Practice Standard 393 (Filter Strip).

Select sites where soils are suitable for pond construction

Soil is the material used to construct embankments that contain water, and when properly formed and compacted, provide a barrier to the movement of water from the pond as seepage. Negative environmental impacts associated with ponds constructed in unsuitable soils are caused by excessive seepage of pond water, excessive water level fl uctuation that requires greater surface or groundwater withdrawals to maintain water level, and sediment loading of receiving waters from embankment erosion. Properly constructed embankments and sufficiently compacted bottom soils will effectively prevent seepage and minimize erosion.

Soil texture is the most important physical characteristic of soils for pond construction. Texture depends on the sizes and shapes of particles and the distribution of those varied particles in a soil. Soils contain coarse-grained particles (e.g., gravel and sand), fi ne-grained particles (i.e., silts and clays), organic matter, and water. Texture class and grain-size distribution should be determined prior to pond construction.

A good soil for construction of embankments is said to be well graded, meaning that it has a wide range of particle sizes that are capable of good compaction. Previously, Hajek and Boyd (1994) recommended that soils for pond construction have a clay content >35%, although this recommendation has since been revised (Boyd et al. 2002). Soils with a high clay content can be difficult to work with machinery because they are highly plastic and cohesive. A clay content of 15% is preferred although soils with a clay content as low as 5 to 10% can be used to construct embankments if soils are well graded.

Water content has a profound impact on the engineering properties of fi ne-grained soils. As water is added to a soil, it transitions from solid to semisolid to plastic to liquid states. Soils with a water content greater than the liquid limit should be dried prior to construc-tion of embankments. Conversely, soils may be too dry to compact properly and water may need to be added.

The depth, thickness, composition, and extent of the impervious base layer should be determined. Often this corresponds to the B soil horizon, which should be 30 to 45 cm thick throughout the impounded area. Other edaphic factors affecting seepage include the thickness of the organic layer, which should be <50 cm, and the depth to hardpan or bedrock, which should be >150 cm (Hajek and Boyd 1994). Pond areas with sand lenses or former stream channels can be sealed by replacement with clay, although such measures can be expensive.

Soil texture, water content, and other soil properties should be evaluated in soil samples from cores and test pits throughout the construction area. For watershed ponds, soil inves-tigations should be done along the centerline of the dam, at the location of the spillway, and in the planned borrow area.

POND CONSTRUCTION

Every effort can be made to select a suitable site, but if ponds are poorly constructed, the inherent advantages of the site cannot be fully realized and the potential for negative environmental impacts increases. Poorly constructed embankments and inadequately com-pacted pond bottoms can result in excessive seepage losses that can contaminate ground-water aquifers or surface water supplies. Embankments that are not properly constructed are also subject to erosion that can contaminate surface water supplies with suspended sediment. Reducing seepage losses with proper pond construction can reduce the environ-mental impact of lost water, reduce water level fl uctuation, and reduce the requirement for water from external sources.

Better Management Practices for Pond Construction

Prepare the site prior to pond construction

Before construction, land should be cleared of all vegetation, roots, stumps, and other obstructions. The upper 5 to 10 cm of topsoil should be removed and stockpiled for use after levee construction is complete to cover embankments in preparation for seeding and other erosion-protection measures. This stockpiled soil should be surrounded by a sedi-ment curtain or hay bales to prevent erosion during pond construction. Any coarse-grained pervious soils, such as can be found in existing stream channels, should be removed, replaced with well-graded soils, and compacted. The soil surface upon which embank-ments are to be constructed should be scarifi ed to facilitate the creation of a good bond between existing soil and material added to form embankments.

Construct ponds according to specific design criteria

The long axis of ponds should be oriented to avoid erosion of levees during the period of strongest winds. In the southeast United States, the strongest winds occur during winter, and a north-south orientation should be avoided. For ponds 4 ha or less, the long axis should be oriented parallel with the prevailing wind direction. For ponds larger than 4 ha, the long axis should be oriented perpendicular to the prevailing wind direction.

Pond bottoms should be smooth and fl at to allow effi cient seine-harvesting of fish. The pond bottom should be constructed with a slope toward the drain of 0.2 to 0.3%. The bottom should be tilled with a disc, leveled, and compacted.

Embankment and excavated ponds for most finfish should be 1 to 2 m deep; crayfish ponds should be 30 to 60 cm deep. Ponds may contain a harvest basin 30 to 45 cm deeper than the pond bottom, or, in deeper ponds, a harvest bench that will allow workers to move easily in the pond to harvest fish.

Embankment top widths should be 6 to 7.5 m on primary levees and 5 to 6 m on sec-ondary levees. The minimum inside slope should be 3 : 1 with a maximum inside slope of 4 : 1; outside slopes should be 3 : 1. The minimum freeboard should be 30 cm for ponds <4 ha and 60 cm for ponds 4 to 8 ha. Freeboard height should be increased above the minimum by the following amounts based on fetch:

Fetch (m)	*Freeboard Increase (cm)*
<100	10
100–200	30
200–400	40
400–1600	60

Embankments for watershed ponds should include a cutoff core trench in the embank-ment foundation. Soil in the cutoff trench should be tightly compacted. The core trench should extend to the height of the permanent water elevation.

Watershed ponds should not have watershed areas larger than necessary to keep ponds full, because excessively large watersheds increase runoff into ponds and result in high discharge. The watershed area required to maintain water level is much greater in arid areas than humid areas. The USDA Natural Resources Conservation Service recommends a watershed area of 13.1 m^2 for every m^3 of storage or about 20 ha for a 1 ha pond that is 1.5 m deep. For watershed ponds in west-central Alabama, about 10 ha of land is required to maintain the water level for 1 ha of pond water surface (Yoo and Boyd 1994). The ratio between annual runoff volume and pond volume should not exceed 3 : 1 to 4 : 1 to avoid excessive fl ushing (Hajek and Boyd 1994). Pond design will vary with climate, topo-graphy, soil type, and plant cover in the watershed.

Ponds should be constructed with properly designed drainage structures. Drainpipes should be 20 cm in diameter or be able to handle the runoff from a 10-year, 24-hour storm, whichever is greater. Most commercial catfish ponds have 25-cm diameter drainpipes. Drainpipes should be installed by excavating a trench after embankment construction is complete. Drainpipes should extend well past the outside toe of the levee to minimize embankment erosion during pond draining (Fig. 6.6). In watershed ponds, anti-seep collars should be added to drainpipes. Soil placed around the pipe barrel should be thoroughly compacted with vibratory tampers up to 60 cm above the pipe. Risers that control water level can be placed inside or outside pond embankments. Trash racks or screens should be placed on internal pipe risers unless the watershed does not contain debris that would clog the outlet.

Spillways constructed for watershed ponds where runoff is the primary water source should be able to accommodate the runoff from a 10-year, 24-hour storm event. Spillways should be fl at at the spillway entrance, allowing for a shallow flow of water from the pond, and vegetated to limit erosion. Beyond the fl at entrance, the spillway should be constructed with a drop to prevent wild fish from entering the pond. In the United States, the local USDA Natural Resources Conservation Service has developed conservation practice standards for pond design and construction and can provide specific technical guidance in these areas. Consult NRCS Conservation Practice Standards 378 (Pond), 397 (Aquaculture Ponds), and 402 (Dam) for additional guidance (www.nrcs.usda.gov/technical/standards/nhcp.html). Specifi c guidance regarding large earthen dam construction can be found in NRCS Technical Release 60 (Earth Dams and Reservoirs).

Compact pond embankments and bottoms properly

Compaction of well-graded soils increases close packing of soil particles, reduces air voids (porosity), and increases the dry density of soil. The objectives of compaction are to decrease future settling, increase shear strength, and—most importantly for aquaculture ponds—decrease permeability. When properly compacted, well-graded soils will result in embankments that are stable and resist erosion and a pond bottom with low hydraulic conductivity to limit seepage.

Compaction is a function of soil dry density, water content, compactive effort, and soil type. Perhaps most importantly, effective compaction of clay soils depends on water content. When water content is low, soils are hard and fi rm; when water content is high, soils are plastic and difficult to compact. For soils with high clay content, the best compaction occurs at a water content slightly less than the plastic limit. Optimum compaction can be achieved with a water content of 10 to 20%, although 10 to 15% is best.

Clays require greater compactive effort because soils are cohesive. Thus, embankments formed with these soils may not be stable or have a good weight-bearing capacity when wet. Swelling of clays is greater and soils are more rigid and stronger when compacted on the dry side of the optimum water content than those compacted on the wet side of the optimum water content. The main reason for these observations is that the particle struc-ture of compacted soil tends to be more random on the dry side, whereas particles tend to be more oriented, or parallel, on the wet side.

Well-graded soils can be worked with a combination of kneading and static weight compaction. For fi ne-grained soils, kneading compacters such as sheepsfoot rollers or tamping-foot rollers are best. Rubber-tired (pneumatic) rollers are suitable, but steel drum rollers are not.

The compaction provided by the dirt pans used to scrape soil from borrow areas and deliver it to embankments is acceptable. Soil should be placed in 15 to 20 cm (6 to 8 inch) layers (or *lifts*) and compacted. Depending on soil type, 2 or 3 passes with a tamping foot roller, or 4 to 6 passes with a sheepsfoot roller are sufficient to compact soil. A settlement allowance of 5 to 10% should be considered during pond construction to achieve the design embankment height.

Control erosion during and after pond construction

Construction of pond aquaculture facilities involves dramatic disturbance of the original landscape. Large areas are laid bare as soil is moved to construct embankments, roads, and support infrastructure. Unless denuded areas are protected, rainfall and runoff will erode the bare soil and contribute to downstream eutrophication and cause sedimentation in receiving streams. Erosion control involves using vegetation, stone, gravel, or other structural practices to protect the soil surface from the impact of rainfall and flowing water.

Protect pond embankments from erosion

Erosion of embankments is caused by rainfall and the scouring action of wind-driven waves inside the pond. Erosion that occurs on the inside slopes of the embankment does not have an immediate environmental impact, but sedimentation of eroded soils reduces the useful life of a pond, requiring more frequent renovation. Water released during reno-vation can have a negative environmental impact if suspended sediment loads are high. Erosion that occurs on the outside slopes of the embankment can enter receiving waters directly or accumulate in the farm drainage system, depending on pond location relative to receiving streams and confi guration of the drainage system. In both cases, it is important to protect embankments from erosion as quickly as possible after construction. In a study of sedimentation in commercial catfish ponds, the greatest sedimentation rates occurred within the fi rst year of pond construction (Steeby et al. 2004), suggesting that erosion of unprotected embankments is a major contributor to initial sedimentation.

Embankments can be protected by seeding the embankment with herbaceous vegetation. Selected plants should be well-adapted to the region and resistant to diseases or insect pests. A perennial, low-statured grass with fl exible stems and a rhizomatous, spreading root system, such as Bermuda or centipede grass in the southeast United States, is most appropriate.

Following construction of embankments, any topsoil that was removed in preparation for pond construction should be spread over embankment tops and slopes and tilled. If necessary, soil should be limed and fertilized in association with seeding. The site should be protected from grazing or from compaction by vehicular traffi c until veg-etation is established. Once vegetation is fi rmly established, embankments should be mowed regularly during the growing season to prevent tree growth. Trees can compromise embankment integrity because roots provide channels for water loss. Burrowing animals should not be allowed to become established and should be trapped or killed humanely.

The effect of erosion on internal embankment slopes caused by wind-driven waves can be minimized by a combination of good construction practices and in-place protection measures. Pond orientation should consider prevailing wind direction and speed, and pond size. The minimum freeboard height is 30 cm, with additional freeboard depending on pond fetch (see above).

Embankments should be constructed with 3 : 1 slopes and soil should be properly compacted during construction. Riprap is not normally placed inside aquaculture pond levees, but some areas prone to excessive scouring may be protected by the addition of erosion-resistant material at the pond waterline. However, riprap should be placed on the outlet side of pond drains to prevent excessive scouring during pond draining. Additional information can be found in the USDA-NRCS Conservation Practice Standards 322 (Channel Bank Vegetation), 580 (Streambank and Shoreline Protection), and Technical Releases 56 (A Guide for the Design and Layout of Vegetative Wave Protection for Earth Dam Embankments) and 69 (Riprap for Slope Protection Against Wave Action). For protection of exterior embankment slopes and drainage systems from erosion, additional guidance is provided in NRCS Conservation Practice Standard 412 (Grassed Waterway).

Construct drainage ditches to minimize erosion

Ditches should be sufficiently large and of proper cross section and slope to convey farm discharges without the occurrence of excessive current that can cause bottom scouring and erosion of the sides. Controlling the flow rate of effl uent from ponds can minimize ditch erosion. Extending drainpipes beyond the toes of pond embankments reduces erosion at the point of discharge. In addition, riprap can be placed on the embankment opposite the point of discharge to defl ect scouring flows.

Eliminate steep slopes on farm roads and cover roads with gravel

Steep grades should be avoided on farm roads, and roads should follow contours whenever possible. A 2- to 6-inch layer of gravel should be applied on farm roads.

Avoid leaving ponds drained in winter, and close valves once ponds are drained

Rain falling on drained or partially constructed ponds can cause serious erosion of internal embankments and pond bottoms, sedimentation in deeper areas of ponds, and loss of suspended solids through the open drain. Refilling ponds promptly and keeping drain valves closed in empty ponds are practices that can protect the pond infrastructure and reduce suspended solids loads to farm ditches and fi nally to receiving streams. If rains occur during construction or renovation, accumulated rainwater can be held for 1 to 2 days prior to discharge to allow any suspended solids to settle before discharge.

POND RENOVATION

Over time, erosion of pond embankments will increase the risk of embankment failure. Soil eroded from internal embankments will accumulate as sediment on the pond bottom, decreasing pond depth and harvest efficiency. Ponds should be renovated before embank-ments can no longer support vehicular traffi c safely. Aquaculture pond renovation tech-niques are reviewed by Steeby et al. (1998). The basic approach to renovation is to dry pond sediments sufficiently to allow bulldozers to enter the pond to replace material eroded from embankments. Replacement of material from the pond bottom to embankments should restore the original levee top width and cross-sectional profi le with interior slopes of at least 3 : 1.

In a technique outlined by Steeby et al. (1998), pond renovation is accomplished in phases. Initially, after ponds are drained, a perimeter trench is excavated with a dragline to allow water to drain from accumulated sediment. Next, a bulldozer is used to cut chan-nels into the accumulated sediment and pile it into rows parallel with the long axis of the pond. Further drainage can occur through the channels. Once the sediment has dried to the proper moisture content (<25%), the material piled in rows can be pushed onto the existing embankments and compacted. Pond embankments should be revegetated as quickly as possible to extend useful pond life.

The major environmental impact of pond renovation is associated with the discharge of sediment from within the pond that is produced while the pond is drained and heavy equipment is operated. Reducing the environmental impact of sediments produced during pond renovation represents a compromise between timely renovation and minimizing sediment release. Specifi cally, water that drains through the perimeter trench and sediment channels will reduce the water content of accumulated sediment, but may carry elevated sediment loads.

Better Management Practices for Pond Renovation

Use sediment from within the pond to repair embankments rather than disposing of it outside of ponds

Ponds are unlike other aquaculture systems because solids derived directly from aquatic animal fecal matter are not the major constituent of the material that accumulates as sediment in the system. Fecal matter produced by animals is quickly decomposed in ponds and the nutrients released in the mineralization process support the growth of aquatic plants—usually phytoplankton—which constitute the majority of suspended solids in most ponds. The organic matter that settles to the bottom rapidly decomposes, especially in warmwater aquaculture ponds, and annual rates of organic matter decomposition nearly match rates of organic matter deposition (Box 6.1).

Nearly all of the sediment that accumulates in earthen aquaculture ponds consists of inorganic solids derived from erosion of pond embankments or erosion on the watershed. The sediment deposition rate can be quite high, especially in large ponds with long fetches for wave generation by wind. For example, Steeby et al. (2004) showed that sediment accumulates in large commercial catfish ponds as a function of pond age according to this equation: accumulation (cm) $= 12.51(x)^{0.46}$, where x = years of continuous pond use. For a pond in continuous use for 9 years (the average for channel catfish foodfish production ponds; USDA-APHIS 2003), an average of 30.4 cm of sediment accumulates in pond bottoms. Most sediment is deposited in deeper areas of the pond, typically the end with the drainage structure. This sediment is a valuable resource that is needed to restore proper pond morphology when the pond is repaired. Solids also accumulate in settling basins, constructed wetlands, or vegetated ditches that capture solids during pond draining. Accumulated solids will eventually reduce the volume of the basin or ditch, affecting hydraulic retention time and the efficiency of solids removal. These solids should be removed or compacted periodically to maintain treatment efficiency.

Reduce discharge of sediments during renovation

During renovation activities, pond drains should remain closed. After activities cease, sediment should be allowed to settle for 1 to 2 days before drains are reopened to release water. Water should be allowed to accumulate in the pond during rainfall events because these can cause sediment suspension and erosion from the exposed pond bottom. Water can be released by opening the drain after a period of settling following a rainfall event.

During pond renovation, excavate to increase operational depth

Increasing depth during renovation can increase water storage, permit greater fl uctuation in water level without compromising aquaculture production, and reduce the volume of effl uent. Using deeper ponds also extends the useful life of ponds by allowing a longer period for sediment accumulation before ponds become too shallow and require renovation. Increasing depth will, however, substantially increase the cost of pond construction.

Properly dispose of solids removed from settling basins

This will be a site-specific practice. Accumulated solids are unlikely to contain toxic concentrations of metals or other chemicals and therefore can be used without concern as fill material.

Overflow Effluents

Ponds have two types of effl uents. One type occurs when water overflows from ponds when rainfall exceeds pond storage capacity or when water is pumped into the pond for intentional water exchange (called *fl ushing*). The other type occurs when water is dis-charged intentionally when ponds are drained. The two effl uents differ in quality, volume, and discharge frequency. Overflow effl uents are discussed here; draining effl uents are discussed in the next section.

Concentrated animal feeding operations (CAFOs), such as cattle feedlots, must be designed to contain all animal waste plus runoff from storm events. Depending on the size, location, and design of the facility, runoff must be contained from a 24-hour, 25-year rainfall event or from all storms, even those exceeding the hypothetical 24-hour, 25-year event (Federal Register 2003). Discharge from CAFOs is prevented by constructing retention ponds to retain wastewater and runoff. Water in the retention structure is either lost by evaporation or applied to land as part of a facility-specific nutrient management plan.

Most pond aquaculture in the United States is conducted in the southeast—a region with abundant rainfall. Using retention ponds to prevent discharge from aquaculture ponds does not appear generally feasible in that region. Most aquaculture farms occupy larger land areas than CAFOs and, consequently, the amount of water discharged from ponds during storm overflow may be much greater than that discharged from CAFOs. For example, Boyd and Queiroz (2001) calculated that 1.53 ha of retention pond would be needed to retain the overflow from a 1-ha embankment pond after a 25-year rainfall event in Alabama. For a typical watershed pond, 11 ha of retention pond would be needed for each hectare of production pond. Therefore, aquaculture ponds in the southeast and other areas where rainfall is relatively abundant cannot be operated profitably without effl uent because of the costs of committing more land area to retention ponds than for production. Furthermore, the reduction in potential pollutant loading from the construction and oper-ation of retention ponds to treat overflow effl uents from aquaculture ponds is far less than those used to treat runoff from CAFOs.

When rainfall exceeds pond storage capacity and causes ponds to overflow, the rate of water exchange during the event is usually low because the volume of most ponds is large relative to the volume of water added (except, however, for small ponds or prolonged, heavy rains). Therefore, the quality of most overflow effl uents is similar to or dilute com-pared with water in the pond prior to the event causing overflow. Solids in overflow effl u-ent are principally phytoplankton, phytoplankton-derived detritus, and fi nely divided clay derived from pond bank and watershed erosion. Since ponds act as their own settling basins and solids that settle rapidly are constantly removed from pond water, the solids that remain in overflow effl uent have poor settling characteristics and are difficult to remove in postdischarge treatment. Reducing mass discharge is best accomplished by reducing effl uent volume, reducing erosion, and using effi cient feeding practices to reduce internal waste loading.

Better Management Practices for Overflow Effl uents

Reduce or eliminate intentional water exchange

Water exchange, or fl ushing, was commonly used in the past to improve pond water quality. Water exchange was thought to remove toxic fish metabolites and excess phyto-plankton and add oxygen or desirable natural foods. Many times, water exchange was used as a desperation measure during fish kills when other measures failed to resolve the problem. As research progressed and the various causes of fish kills in aquaculture ponds were identifi ed, there was less justifi cation for water exchange as a pond management practice.

If freshwater ponds are managed within their assimilative capacity for waste nutrients, there should be few (if any) times when water exchange is needed to improve water quality. That is, in fact, part of the definition of pond aquaculture. When waste loading exceeds the capacity of physical, chemical, and biological processes to eliminate or trans-form the wastes, water exchange is needed to prevent water quality deterioration. At that point, the system transitions from a pond to a flow-through facility.

For large ponds, such as those used in channel catfish aquaculture, incoming water is greatly diluted by pond water, and it is unlikely that enough water can be exchanged in a sufficiently short period of time to have a beneficial effect during acute water quality crises. Research (McGee and Boyd 1983) and practical experience have shown that routine water exchange at rates possible in large catfish culture ponds (less than 5% of total pond volume daily) has little or no effect on pond water quality or fish production.

Water exchange is, however, widely practiced as a dissolved oxygen management tool in crayfish culture, where as many as 7 to 9 pond volumes may be exchanged over a production season (Avery and Lorio 1999). Flushing ponds is commonly practiced when crayfish ponds are fi rst fl ooded in the fall, when the decomposition of abundant vegetation places a substantial demand on dissolved oxygen, leading to critically low concentrations. In this situation, ponds should be filled to 30 cm depth, drained to a depth of 10 to 15 cm and then refilled with new water to 30 cm depth (Lutz et al. 2003). When temperature declines and dissolved oxygen concentration improves, ponds can be filled to normal operating depths (35 to 45 cm). In this way, ponds are drained and filled at separate times, improving the effectiveness of water exchange.

Water exchange is also considered an essential part of the fi nal phase of hybrid striped bass culture, although there has been no systematic research to support the practice. Intentional water exchange is rare in baitfish and ornamental fish culture because fl ushing with new water reduces the abundance of natural foods, which are an important source of nourishment for those fish.

Displaced pond water represents a pollution load to receiving waters and consumptive use of a valuable resource. Furthermore, pumping requires energy and adds to the cost of growing the aquaculture crop. Accordingly, high rates of water exchange should not be used unless absolutely necessary. If water exchange is practiced as a dissolved oxygen management tool, pond aeration or mixing should be considered as a preferable option. Where technically feasible, discharged water should be circulated between a treatment or reservoir pond and production ponds.

Manage ponds to capture rainfall

Although most aquaculture ponds cannot be operated economically without some water being discharged during unusual precipitation events, the volume of water discharged can be greatly reduced by keeping the pond water level below the level of the drain so that rainfall is captured rather than allowed to overflow. This method of reducing effl uents, originally modeled as a water-conservation practice for catfish ponds (Pote et al. 1988; Pote and Wax 1993), is known as the *drop-fill scheme.*

In drop-fill schemes, pumped water is not added until evaporation and seepage cause the pond water level to fall to a predetermined level below the level of the pond overflow device. At that point water is added to replace losses, but the pond is not refilled com-pletely. For example, a *10–4 drop-fill scheme* means that pumped water would not be added until the pond water level falls 10 inches (25 cm) below the level of the overflow device. Then, 4 inches (10 cm) of water would be added, bringing the water level to 6 inches (15 cm) below the overflow level. This 6 inches (15 cm) of storage will retain considerable rainfall, thereby reducing overflow and pollutant mass discharge, as well as offsetting the need for pumped water (Tucker et al. 1996) (Box 6.2).

Another approach to reducing effl uent volume is based on the use of some ponds on a farm for both fish production and water storage. The combined production-storage pond is constructed 0.25 to 1 m deeper than typical production ponds to provide additional volume for storage of rainfall.

The production-storage pond is linked via culverts to 1 to 3 production ponds so that overflow from all ponds in the linked system drains into the production-storage pond. Overflow occurs only when the storage capacity of the linked system is exceeded. Mathematical modeling using a 26-year climatological record for northwest Mississippi showed that the effl uent discharge from linked ponds can be reduced by 40 to 90% (depending on rainfall) relative to single ponds refilled to the top of the overflow device every time the water level drops 7.5 cm (Cathcart et al. 1999).

Optimize the ratio of watershed-to-pond area

Watershed ponds should not have watershed areas larger than necessary to keep ponds full because excessively large watersheds increase runoff into ponds and result in high discharge. Optimal pond design will vary with climate, soil type, and watershed use. Consult representatives of the local offi ce of the USDA Natural Resources Conservation Service for assistance with pond design. See the specific recommendations in the section on pond construction above.

Use drains with surface-water intakes

Although shallow embankment ponds temporally stratify during the day, convective heat loss at night causes destratifi cation and mixing of the water. Water mixing by wind and aeration also counteracts thermal stratifi cation. The quality of water in embankment ponds is, therefore relatively homogeneous with depth, and except for brief periods at the begin-ning and end of pond draining, effl uent quality does not differ significantly if waters are discharged from either the surface or bottom of the pond. However, deeper ponds may stratify for long periods during warmer months, and water quality can vary from the surface to the bottom. Bottom water has lower dissolved oxygen concentrations and higher dissolved nutrient and suspended solids concentrations. The quality of effl uent discharged as overflow from the surface may be better than that discharged from the bottom. Pipe sleeves, dam board confi gurations, or other structures that allow the withdrawal of water from deeper layers should not be used. In ponds managed with intentional water exchange, water can be discharged from the pond bottom if it is recirculated or treated in a reservoir pond or wetland prior to discharge.

Control erosion on pond watersheds and pond levees

Vegetative cover will reduce erosion of watershed soils and the loading of suspended solids to ponds in runoff. In watersheds used for grazing, livestock stocking rates appro-priate for the forage base will allow maintenance of sufficient plant cover. In degraded watersheds or those subject to erosion, streamside management zones (SMZs) or buffer strips adjacent to water supply streams should be maintained. Seed pond levees with a cover crop, such as Bermuda or centipede grass, shortly after pond construction. Planting a cover crop will limit sedimentation of the pond, thereby extending the interval before renovation is required.

Divert excess runoff from large watersheds away from ponds

Runoff and supply streams can be diverted by constructing temporary low-elevation embankments or ditches running across slope on watershed areas adjacent to ponds. This practice can divert turbid runoff where it is not possible to provide erosion control on watersheds. Diversion can extend the production interval between draining and renovation in watershed ponds by minimizing reduction in pond depth caused by deposition of sedi-ment derived from turbid runoff. Less water passing through ponds will reduce erosion of embankments and farm ditches, and the output of suspended solids will be less.

Whenever possible, prevent water discharge from ponds during fish kills

If the fish kill was caused by deleterious water quality, preventing water discharge during fish kills will limit the release of potentially poor-quality water. Furthermore, preventing water discharge from ponds during fish kills will limit the release of potential fish pathogens. Drainage structures can be temporarily plugged or blocked to prevent overflow.

Use effl uents to irrigate terrestrial crops

Under certain conditions, the water discharged as intentional overflow or from pond drain-ing may have value as irrigation water for agronomic crops. The primary goal of integrat-ing aquaculture and terrestrial agriculture would be to make productive use of pond effl uents by supplementing the water supply for irrigation. Nutrients in pond effl uent might also be beneficial to the crop (Box 6.3). However, there are a number of practical consid-erations that severely restrict the utility of this seemingly attractive management practice. The primary impediment to using pond effluents for crop irrigation is that crop water requirements seldom correspond to the availability of pond effl uents. Peak water demand for irrigation occurs during dry periods when there is little or no overflow from ponds. Also, many fish farmers object to draining ponds during the summer growing season because they lose valuable production time when ponds are empty. Most farmers prefer to drain ponds in autumn or winter, during which time there is little demand for irrigation water. Because irrigation demands do not correspond to water availability from rainfall overflow or intentional pond draining, the only way to provide water for irrigation is to intentionally pump water into the pond to generate overflow.

In some types of irrigation (such as fl ooding of rice fields) water is needed quickly, in relatively large volumes, and at a specific time—which may or may not correspond to the availability of water from fish ponds. For example, a crop of rice requires about 1 to 1.5 m of water, which corresponds to nearly the entire volume of most fish ponds for an equiv-alent area of rice. Low rates of water exchange do not improve water quality in fish ponds (McGee and Boyd 1983), so using pond water for irrigation would not benefit fish produc-tion. Further, the need for large water volumes delivered quickly means that gravity flow from pond discharge devices may not be sufficient to provide the flow rates that are needed. Rapid delivery of large water volumes would require installation of large pumps and water distribution systems to convey water where it is to be applied in the required amounts. The high suspended solids concentration of some aquaculture pond effl uents can clog the water distribution orifi ces of some kinds of irrigation equipment. Thus, use of pond effl u-ents with some types of irrigation equipment would require fi ltration, a process that would remove a significant proportion of the nutrients contained in the effl uent.

EFFL UENTS FROM POND DRAINING

The quality, volume, and frequency of water discharged when ponds are drained is dif-ferent from that discharged as overflow following rains or from intentional water exchange. Overflow effl uent is unpredictable and discharge volume and frequency varies depending upon weather. Pond draining, on the other hand, is predictable and usually varies with the culture cycle. When ponds are drained, there are periods (often quite brief) when solids from the pond bottom are discharged along with pond water. Accordingly, water dis-charged during pond draining may temporarily have much higher concentrations of solids, nutrients, and organic matter than overflow effl uents.

Frequency of pond draining (and, therefore, annualized effl uent volume) depends pri-marily on the species or phase of culture. For some species, such as ornamentals and baitfish, crops are completely harvested annually (or more frequently) and ponds are drained to facilitate harvest and prepare the pond for the next crop. Ponds used to grow crayfish and juvenile phases of channel catfish and hybrid striped bass also are drained annually. In contrast, ponds used to grow food-sized channel catfish may be operated without draining for many years.

Pond hydrological type also affects discharge frequency and effl uent quality. Embankment ponds used to grow food-sized channel catfish are relatively shallow (1- to 2-m deep) and there is no need to partially drain ponds to facilitate fish harvest. Ponds may be operated without draining for many years by selectively removing market-sized fish and restocking the pond with fi ngerlings. When the pond must be drained after years of use, a seine is pulled through the pond one or more times to capture fish before draining. In contrast, deep watershed ponds are difficult to harvest without partially drawing down the water level. Watershed ponds may need to be partially or completely drained each year.

Most embankment ponds for channel catfish are equipped with fixed drains that extend from the deepest part of the pond through the embankment. A screened extension is fi tted to the end of the pipe (usually 25 to 30 cm in diameter) inside the pond, with a riser and an "alfalfa" valve on the outside. The height of the riser determines pond depth (Fig. 6.7). Draining is initiated by removing the riser extension. Then, after the water level in the pond has dropped at least half way, the valve at the bottom of the drain is opened and the remaining water in the pond is drained. When draining is initiated, shear forces generated by water moving into the drainpipe scour the pond bottom around the entrance to the drainpipe. The initial fl ush of water discharged, therefore, consists of pond water and a slurry of sediment that has accumulated over the screen inside the pond. During this fi rst fl ush of sediment-laden water, concentrations of solids, organic matter, and nutrients can be very high. Depending on the depth of sediment around the drain, the effl uent clears in 5 to 30 minutes and all water subsequently discharged is pond water with little or no material derived from the sediment (Hargreaves et al. 2005a). Although concentrations of potential pollutants can be high in the fi rst-fl ush effl uent, the duration of elevated concen-tration is relatively brief and, therefore, the proportion of pond volume (and total pollutant mass) discharged during that period is small (Hargreaves et al. 2005b).

Some watershed ponds are too deep to allow fish harvest unless the pond is partially drained before the seining operation begins. In those ponds, effl uent quality is more or less constant during discharge of the initial 80 to 90% of the pond volume (the draining phase). Following discharge of most of the pond water, fish are harvested by repeatedly seining the water remaining in the pond. Disturbance of the sediment by harvest activities and fish movements in shallow water causes nutrient, organic matter, and suspended solids concentrations to increase dramatically during the last 10 to 20% of discharge volume (Boyd 1978; Schwartz and Boyd 1994).

Mass discharge from pond draining can be minimized by reducing the volume of drain-ing effl uent, primarily by extending the interval between pond drainings. If possible, this should be the fi rst approach. Next, the water quality of the effl uent from pond draining can be improved by practices that vary depending on hydrological pond type. The least practical approaches are those where effl uent is treated after discharge.

Better Management Practices for Effl uents from Pond Draining

Reuse water for multiple crops

It is impossible to conduct pond aquaculture without occasionally draining the pond. Most commonly, ponds are drained to facilitate complete harvest of the aquaculture crop. Production schemes have, however, been developed for some species that allow multiple crops to be grown without draining the pond. Using such systems dramatically reduces effl uent volume. Therefore, whenever possible, ponds should be constructed with uniform, relatively shallow (1 to 2 m) bottoms that allow harvest without draining. In multiyear production systems, such as that used to grow food-sized channel catfish, ponds are drained only when renovation is needed or when there is need to adjust fish inventory by completely harvesting the crop. Wind-driven waves, acting over several years, can erode levees to the point where levee integrity is compromised. Material eroded from levees also accumulates as sediment on the pond bottom, making the pond bottom uneven and difficult to seine-harvest. It is also difficult to maintain accurate records of fish inventory in multiyear production systems, and farmers sometimes harvest and drain ponds simply to "zero-out" the inventory and start a new production cycle. The interval between pond drainings can therefore be increased by managing ponds to reduce erosion of earthwork and by improving inventory control through effi cient harvest and better record keeping.

Using ponds for several years without draining and refilling is possible, in large part, because natural microbial and physicochemical processes continually remove nutrients and organic matter from pond water. The rate at which these processes act is such that fish ponds (at least in the southeastern United States) can be used for many years without significant long-term accumulation of nutrients and organic matter in the water column or sediment, despite large inputs of metabolic waste resulting from fish feeding practices (Box 6.4).

Reuse or recirculate water that is drained from ponds

Rather than draining ponds for harvest, it may be possible to pump water to adjacent ponds where it can be reused in the same or other ponds. To facilitate this practice, ponds can be constructed with higher levees than normal or water levels can be maintained with more freeboard to provide greater storage capacity for water pumped from a nearby pond.

Water from one pond can be transferred to another with a low-head lift pump and then, where topography permits, returned to the empty pond by siphon. Alternatively, drained water can be lifted to a large reservoir pond for treatment prior to recirculation to culture ponds by gravity. If topography of the landscape allows, ponds can be constructed in such a way to allow draining of one pond into adjacent, lower-elevation ponds.

Reusing pond effl uent is commonly practiced on baitfish farms in the Grand Prairie region of eastern Arkansas. Water reuse practices have been widely adopted because of the dwindling groundwater supply, the lower cost of pumping water from pond to pond compared with pumping water from deep wells, and a desire to not waste the "old water" that contains abundant natural foods for baitfish.

Drain water from the pond surface

The quality of pond surface waters is often better than that near the pond bottom. Using drainage structures that allow control over the depth from which water is discharged con-trols, to some extent, the quality of water discharged. Examples of fl exible-level drainage structures include swivel drains (Fig. 6.7) and drainage monks managed with progressive removal of dam boards as ponds drain. These drainage structures allow the discharge of pond water in stages. The last 10 to 20% of water can be held until solids settle and then discharged.

Allow solids to settle before discharging water

Some watershed ponds must be partially drained to facilitate fish harvest. In those instances, water should not be discharged during seining to avoid the discharge of sediment sus-pended by harvest activities. After seining, the last 10 to 20% of water can be held without discharge to allow solids to settle. Holding water for 2 or 3 days after seining can greatly reduce the discharge of solids, organic matter, and nutrients (Schwartz and Boyd 1994). Holding water for more than 3 days can result in an increase in solids concentration as phytoplankton blooms develop in response to the nutrients released during harvesting.

Treat pond effl uents in constructed wetlands prior to discharge

Using wetlands to treat wastewaters is based on removal of nutrients and solids as the water is slowly passed through a shallow (30-cm deep), vegetated impoundment. Nutrients are assimilated by wetland plants, removed by physicochemical processes, such as precipitation and adsorption reactions in the soil, and transformed and removed by biological reactions associated with the vast surface area provided by plant roots and above-ground plant biomass. Solids are removed by settling as water slowly passes through the system.

The disadvantage of constructed wetlands for treating wastes from channel catfish ponds is the large area necessary to provide adequate hydraulic retention time when the process is used on large farms. In particular, the relationship between the pond area and wetland area needed for effective effl uent treatment during the high-discharge winter periods is not known. Also, the usefulness of wetlands to treat pond effl uents in colder climates is suspect.

An alternative to using wetlands designed to treat all effl uents from a farm would be to construct a small wetland to treat only the most concentrated effl uents released when ponds are drained in the drier seasons—the time of greatest potential environmental impact because of low rates of dilution by effl uent-receiving streams. In other words, effl uents released during peak discharge would not be treated because of the low nutrient concentra-tion of most of the effl uent volume, the large wetland area needed to provide an effective hydraulic retention time, and the great dilution provided by high receiving stream flows. This approach would minimize the land needed for constructed wetlands and significantly improve effl uent quality during dry periods. However, the overall reduction in mass dis-charge of nutrients and organic matter, and the costs associated with this scaled-down approach, are unknown.

Treat pond effl uents in settling basins prior to discharge

Settling basins are easier to construct and operate than wetlands because they do not require the establishment of plants. More important, Boyd et al. (1998) showed that set-tling basins may be nearly as effective as constructed wetlands at removing potential pollutants from effl uents discharged during pond draining, which often contain high con-centrations of solids dislodged from the pond bottom by fish harvest activities. Sedimentation of synthetic pond effl uents (designed to simulate waters released during the fi nal stages of pond draining) for 8 hours removed more than 75% of suspended solids and total phos-phorus and more than 40% of the 5-day biochemical oxygen demand (BOD_5) and turbid-ity. Removal of these substances is rapid because much of the phosphorus and organic matter in this particular kind of pond effl uent is associated with inorganic suspended solids, primarily suspended soil particles, which settle quickly (Box 6.6).

For watershed ponds that must be partially drained to facilitate fish harvest, the produc-tion pond can be used as its own settling basin. As discussed above, when ponds are drained, the fi nal volume of water may be held in the pond for 2 to 3 days to allow solids to settle before draining completely. Alternatively, the last 20% of water discharged from ponds can be held in farm drainage ditches for settling prior to fi nal discharge.

Release pond effl uents into low-gradient drainage ditches

Low-gradient ditches, particularly those with natural or planted vegetation to impede flows, can function as effective settling basins for rapidly settling solids released during pond draining (Box 6.7)

WATER CONSERVATION

Water conservation is a commendable goal because it helps ensure the sustainability of aquaculture. In pond aquaculture there are additional incentives to conserve water. Water for most pond aquaculture requires pumping, either from underground aquifers or surface waters. Pumping can be expensive, depending on the hydraulic head, distance the water must be moved, and energy or fuel costs. So, water conservation reduces production costs. There is also a direct link between water conservation in pond aquaculture and effl uent volume. Every liter of water that is discharged must be eventually replaced, either by rainfall or pumped water. So, water conservation also reduces effl uent volume.

In many parts of the world, aquaculture must compete for water with other forms of agriculture, industry, and household uses. In arid and drought-prone regions, conservative and intensive water management is the only way pond aquaculture can be conducted. For example, baitfish and catfish aquaculture in the Grand Prairie region of eastern Arkansas and ornamental fish culture in south Florida are severely constrained by lack of suitable groundwater supplies caused by long-term overdrafts by other forms of agriculture (rice and cotton in the case of Arkansas) or urban development (in the case of Florida). In the face of this critical resource constraint, pond aquaculturists in those regions have developed innovative practices to conserve water without significantly affecting production.

Losses of water from ponds include evaporation, transpiration, seepage, overflow, and draining. Little can be done to reduce water lost to evaporative processes, but other losses can be controlled to some degree. Most water conservation practices are addressed in other sections of this chapter and will be mentioned only briefl y here. See the sections "Pond Hydrology" for a discussion of water budgets and water loss processes, "Site Selection" and "Pond Construction" for aspects of siting applicable to reducing seepage losses, and "Overflow Effl uents" and "Effl uents from Pond Draining" for practices that can reduce water lost in effl uents.

Better Management Practices for Water Conservation

Select sites with soils that are suitable for pond construction

Soils vary widely in their ability to retain water. Well-graded soils with a range of particle sizes and a moderate clay content will compact well during construction and have negli-gible seepage. Permeable soils with high sand content or soils with lenses of sand or gravel will seep excessively.

Also, thin soils over fractured bedrock (especially over limestone) can seep excessively. Proper site selection will largely prevent excessive water loss to seepage. Landowners should consult with the local USDA Natural Resources Conservation Service (NRCS) offi ce before building ponds. The NRCS has soil surveys with informa-tion on soil properties and their suitability for pond construction. Core samples from a proposed pond site will help identify areas of high permeability.

Construct ponds properly

Construction practices that can reduce water loss to seepage include proper preparation of the land before construction, proper pond design (which will be site-specific), adequate compaction of soil during pond construction, and installation of properly designed drain-age structures. The USDA Natural Resources Conservation Service has developed practice standards for pond design and construction and can provide specific technical guidance in these areas.

Consult NRCS Conservation Practice Standard 378 (Pond), 397 (Aquaculture Ponds), and 402 (Dam) for additional guidance. Specifi c guidance regarding large earthen dam construction can be found in NRCS Technical Release 60 (Earth Dams and Reservoirs).

Reduce or eliminate intentional water exchange

Water exchange is believed to improve pond water quality by removing toxic fish metab-olites and excess phytoplankton, and adding oxygen or desirable natural foods. Although water exchange is considered essential in some types of pond aquaculture, reduction or elimination of this practice is probably feasible in all cultures. Displaced pond water increases effl uent volume and is a consumptive use of a valuable resource. Pumping also requires energy and adds to the cost of growing the aquaculture crop. If water exchange is practiced as a dissolved oxygen management tool, pond aeration or mixing should be considered as an option. Where technically and economically feasible, discharged water can be circulated between a treatment or reservoir pond and production ponds to conserve water.

Manage ponds to capture rainfall

Ponds can be designed and operated to retain rainfall, which will offset future needs for pumped water. Building and operating ponds with extra freeboard and using the drop-fill scheme to capture water (see the section "Overflow Effl uents") can greatly reduce the need for pumped water (Pote et al. 1988; Pote and Wax 1993; Hargreaves et al. 2001; Tucker and Hargreaves 2006).

Reduce water lost during pond draining

Although it is impossible to conduct pond aquaculture without occasionally draining the pond, production schemes have been developed for some species that allow multiple crops to be grown without draining the pond for many years. These production systems dra-matically reduce effl uent volume.

Therefore, whenever possible, ponds should be con-structed with uniform, relatively shallow (1 to 2 m) bottoms that allow harvest without draining. Additional practices that allow for longer intervals between draining include reducing erosion on the watershed, building deeper ponds (Steeby et al. 2004), and using management practices that allow better tracking of pond inventories.

Reuse or recirculate water that is drained from ponds

Rather than draining ponds for harvest, it may be possible to pump water to adjacent ponds where it can be reused in the same or other ponds. Water from one pond can be transferred to another with a low-head lift pump and then, where topography permits, returned to the empty pond by siphon. Drained water may be lifted to a large reservoir pond for treatment prior to recirculation to culture ponds by gravity.

Use water efficiently by increasing production intensity

Water use in pond aquaculture can be described in absolute terms (volume/time), but a better index is water efficiency (fish production/ water volume). As production intensity increases, water efficiency correspondingly increases. Extensively managed ponds are comparatively wasteful of water because production is low relative to the volume of water used. Intensively managed ponds use water effi ciently, although the quality of water is degraded during culture in comparison to that in extensively managed ponds. Intensifi cation of pond aquaculture can be achieved by increasing stocking density and feeding rates, which requires corresponding increases in aeration, mixing, and occasionally water exchange. Water efficiency can also be increased by reusing water for multiple crops before draining. In areas where water is scarce, conservation and production objectives can be achieved through intensification.

FEEDS AND FEEDING

Although some ponds are fertilized in the early stages of production to stimulate natural food production, most finfish aquaculture in the United States relies on manufactured feeds to promote rapid fish growth. Feeds used for catfish and hybrid striped bass culture are manufactured from high-quality ingredients and formulated to meet all nutritional require-ments of the fish. Nevertheless, even under the best conditions only about 10 to 20% of the carbon, 20 to 30% of the nitrogen, and 25 to 35% of the phosphorus in the feed is removed in fish at harvest. The remainder is excreted by fish into the pond. Feed nutrient assimilation efficiency is even lower in baitfish and ornamental fish culture because these fish often do not eat all the feed offered. Dissolved and particulate nutrients and organic matter excreted by cultured animals and derived from applied feeds constitute the waste load to pond water. Nutrients from uneaten feed and fish metabolic waste products stimu-late the production of large amounts of phytoplankton, which constitute most of the sus-pended solids in pond water.

The total amount of a substance discharged from aquaculture facilities (*mass discharge*) is the product of concentration and total effl uent volume. In "open" systems, such as flow-through and net-pen culture systems, effl uent volume is more or less fixed and mass loading is primarily related to the interactions among feed composition, feeding rate and frequency, and fish bioenergetics. Therefore, one of the best ways to reduce waste mass discharge is to improve feed nutrient assimilation efficiency so that less waste is produced within the system. Although there is a general relationship between nutrient input via feed input and pond water quality (Tucker et al. 1979; Boyd 1985b; Cole and Boyd 1986; Southworth et al. 2006), the relationship between feed input and effl uent mass loading is much less direct than it is in open systems.

Two factors disconnect internal waste loading from waste discharge in pond aquacul-ture. First, most ponds discharge water intermittently and mass discharge is usually more sensitive to changes in pond draining frequency and overflow effl uent volume than it is to changes in pond water quality. The second factor that disconnects waste loading from mass discharge is the cumulative effect of various natural, within-pond processes that transform and remove a large majority of the waste load before water is discharged. These physical, chemical, and biological processes are well known in limnology and wastewater treatment, and are not unique to aquaculture ponds (see Boyd 1985b; Hargreaves 1998; Boyd and Tucker 1998; and Hargreaves and Tucker 2003 for discussions of these processes).

The overall effects of these two factors (discharge volume and internal waste removal) on waste discharge are interrelated because the degree to which the wastes are treated in ponds before discharge depends primarily on hydraulic retention time and water tempera-ture, both of which vary considerably depending on pond hydrological type, culture species and life stage, individual farm management practices, season, and geographic location. For example, most ponds used to raise food-sized catfish are not drained between crops, are operated with no intentional water exchange, and water temperatures are warm throughout much of the year. Under those conditions, well over 90% of the nitrogen, phosphorus, and organic matter added as fish metabolic waste is removed from water prior to discharge (Tucker et al. 1996). Draining ponds between crops and using intentional water exchange will reduce hydraulic retention time and afford less time for natural pro-cesses to remove nutrients and organic matter before water is discharged. Because biological activity within the pond and physical factors such as hydraulic reten-tion time and pond draining frequency have such a large impact on waste mass discharge from ponds, feed management offers less opportunity for control of potential pollution than it does in "open" culture systems. In ponds, reducing waste discharge by reducing effl uent volume is easier and more effective than trying to reduce pollutant concentrations through feed management. Improved feed nutrient utilization in ponds does, however, provide other important benefits. Feed represents the largest single variable cost (typically about half) of aquatic animal production and effi cient use of feeds can improve farm profitability. Also, operating ponds within the assimilative capacity of the pond ecosystem will improve water quality inside the pond and provide a better environment for aquatic animal growth. Feed management is therefore one of the most important aspects of pond aquaculture.

Aquatic animal nutrition and feeding practices are active areas of research, and technol-ogy is constantly evolving. An important research goal is to improve the efficiency of nutrient utilization by the animal under culture, thereby enhancing economic returns and reducing waste production. Because technology is rapidly changing, BMPs for feed man-agement should be fl exible so that newer and better practices can be implemented as they become available.

Better Management Practices for Feeds and Feeding

Use feeding practices that maximize the efficiency of feed use

Most feeds used in commercial aquaculture are manufactured to high quality standards and formulations are based on many years of research into the nutrient requirements of the cultured animal. Nevertheless, poor feeding practices can negate the benefits offered by these feeds. In fact, for channel catfish, hybrid striped bass, and marine shrimp, there is much greater opportunity for reducing waste loading in ponds by improving feeding practices than by making incremental improvements in feed formulation. For example, improving the feed conversion ratio (weight feed fed/fish weight gain) from 2.2 to 1.8 in channel catfish farming results in almost 25% less waste nitrogen added to the pond. Reductions in waste loading of this magnitude are possible by using improved culture practices, but are difficult to achieve simply by improving current feed formulations (but see Box 6.8). Regrettably, much more research has been conducted on nutrient require-ments and feed formulation than on feeding practices, and there is no standard feeding practice for any commercially important species grown in ponds. In that regard, feeding fish in ponds is more skill than science. Developing the skill to feed fish effi ciently takes experience to acquire a feel for how fish appetite, as an aggregate population effect, changes in response to variations in weather and pond water quality.

Feeding fish in ponds is not as simple as feeding terrestrial animals because it is often difficult or impossible to observe underwater feeding activity, especially in turbid ponds.

Also, environmental conditions in ponds change more rapidly than in other aquaculture systems, and those changes cause fish appetite to vary from day to day in a largely unpre-dictable manner. Feeding response can be best gauged when extruded (fl oating) feeds are used. Pelleted (sinking) feeds do not allow observation of fish feeding activity and are generally not recommended for finfish aquaculture in ponds.

Daily feed allowances can be based either on subjective evaluation of fish feeding response or on schedules computed from fish size and biomass, expected feed consump-tion, and water temperature. It is difficult to use scheduled feeding in ponds because fish feeding response varies unpredictably with environmental conditions and because fish biomass may not be accurately known in ponds. Therefore, feeding fish in ponds should be based on subjective evaluation of feeding response rather than schedules based on expected feed consumption (Garrard et al. 1990).

Fish growth is usually best when fish are fed one or more times daily to satiation. On the other hand, nutrient retention improves, up to a point, when ration is restricted. The best compromise between growth and feed conversion for commercial pond aquaculture is attained when fish are fed to just short of satiation (Minton 1978). It is, however, dif-fi cult to judge when fish are fed to that level in ponds because fish appetite varies from day to day and from pond to pond. As such, there is no way to know what will constitute satiation until that point is reached. Experienced culturists who pay close attention to feeding activity can, however, judge with fair accuracy when the point of just short of satiation is reached. Healthy, actively feeding fish should be satiated within 20 to 30 minutes of feeding, and feeding fish more than they can consume in this relatively short period is wasteful and results in poor feed conversion. The previous discussion points out the importance of carefully feeding fish. However, logistical considerations associated with feeding a large number of ponds on large farms often mean that fish are not fed with the care and attention that would result in improved feed conversion and profitability. In this case, it is important to weigh the potential gains associated with improved feed con-version and increased production against the costs associated with additional labor for feeding carefully.

Other examples of effi cient feeding practices include feeding from the upwind side of ponds, broadcasting feed across as large an area as possible, and taking time to observe fish feeding behavior. Feeding rates should be adjusted with season, with lower feeding rates used in winter.

Base feeding regimes on the capacity of the pond to assimilate waste nutrients and organic matter

Water quality deteriorates when the waste load derived from feeding exceeds the capacity of the pond to assimilate waste nutrients and organic matter. Impaired water quality stresses aquatic animals and reduces the efficiency of feed conversion and production. High stocking and feeding rates also lead to effl uents with a greater pollution potential, although the impact of feeding rates on effl uent quality depends on when effl uent is dis-charged relative to the time of maximum feeding rates. Nevertheless, stocking and feeding at profit-maximizing rather than yield-maximizing rates is more effi cient (Losinger et al. 2000) and the likelihood of water pollution by effl uents is less. There is no set value for pond assimilative capacity, which can vary greatly depending on water temperature, pond size and depth, amount of supplemental aeration, water exchange rate, and other factors (Hargreaves and Tucker 2003). Assimilation capacity can be loosely determined empiri-cally by monitoring fish feeding response. A reduced feeding response that cannot be attributed to infectious disease may indicate that pond assimilative capacity has been exceeded.

In practice, maximum sustained feeding rates in ponds are determined by the avail-ability of dissolved oxygen, which in turn is a function of the amount of mechanical aeration. By enhancing dissolved oxygen concentrations, aeration increases the capacity of ponds to assimilate organic matter by aerobic processes. Higher dissolved oxygen concentrations also increase the nitrifi cation rate of ammonia to nitrate, which is then lost from the pond through denitrifi cation in the large volume of anoxic sediment. Aeration and water circulation also affect rates of phosphorus loss from pond water. Formation of oxidized ferric phosphates and phosphorus occluded in ferric oxyhydroxide coatings on soil particles in the surface layers of sediment are important sinks for orthophosphate from the overlying water. In addition, the thin layer of oxidized surface sediment functions as a barrier that prevents the release of orthophosphate from deeper, anaerobic layers of mud into the overlying water (Masuda and Boyd 1994a, b).

In general, long-term daily feeding rates in summer months should not exceed about 30 kg/ha in ponds without mechanical aeration. Long-term average daily feeding rates can be increased to 110 to 140 kg/ha per day in ponds provided with effi cient mechanical aerators at 2.5 to 3.5 kW/ha. Higher amounts of mechanical aeration may allow for higher feeding rates, but these more intensive pond systems may require periodic water exchange or additional energy inputs for continuous mixing to allow nutrient inputs to be sustained at that level. Maximum daily feeding rates of 300 kg/ha are possible with 4.25 to 8.50 kW/ ha aeration and 30% daily water exchange. Obviously, the discharged water represents a potential pollution source that should be treated on-farm before reuse or discharge.

Despite the clear benefits of aeration, excessive aeration and circulation can have del-eterious effects. Erosion of pond bottoms and embankment by strong water currents pro-duced by certain types of aerators can suspend large amounts of particles and potentially increase the suspended solids concentration in pond effl uents. Zones of scouring and deposition create an uneven pond bottom profi le that renders harvesting fish with seine nets difficult and ineffi cient.

Use high-quality feeds

Using high-quality feeds improves feed nutrient retention, thereby reducing amounts of metabolic waste and uneaten feed. Bulk or bagged feeds should be manufactured to contain a minimum amount of fi nes, or dust. Fish feeds are more fragile than feeds for livestock and poultry and will naturally contain 3 to 5% fi nes. Fines should not be fed from bagged feeds and it may be possible to screen fi nes from bulk feeds. Feeds should be stable in water long enough so that pellets remain intact until eaten. The dry matter retention after 1 hour, an index of water stability, should be >90%. Compared with fl oating feeds manu-factured by cooking extrusion, steam-pelleted (sinking) feeds fracture more easily during shipping and handling and disintegrate more rapidly in water. Although more expensive, extruded feeds are therefore recommended for most pond aquaculture situations.

Feeds should be formulated to meet the nutritional requirements of the cultured species and should be formulated using practical ingredients that have high dry matter and protein apparent digestibility. Formulations should be designed to enhance nitrogen and phospho-rus retention efficiency and reduce metabolic waste output (Box 6.8). Feeds should contain sufficient dietary energy to spare dietary protein (amino acids) for tissue synthesis. Feed protein levels should be as low as possible without causing adverse effects on fish growth, health, processing yield, or fillet quality. To a large extent, feed protein requirements refl ect the trophic position of the cultured animal, with higher levels required for carnivo-rous species and lower protein levels for herbivorous or omnivorous species. Generally, higher levels of feed protein are also required for culture of the early life stages of aquatic animals. Available phosphorus levels should be slightly in excess of the dietary require-ments of the cultured species for each life-history stage, and the formulation should be designed to minimize the difference between total feed phosphorus levels and available feed phosphorus levels. Consult a qualifi ed aquatic animal nutritionist or feed manufac-turer for information regarding feed formulation.

Use feeds with the least amount of fishmeal and other animal protein as possible

Cultured fish do not have a requirement for fishmeal, having only a requirement for specific essential amino acids, minerals, and fatty acids. Fishmeal is an excellent source of these nutrients, with amino acid profi les similar to that of the requirements of cultured finfish. Fishmeal is also an excellent source of energy and is highly palatable. The protein and energy in fishmeal is highly digestible, implying that waste loading from fish fed diets with high levels of fishmeal is low. However, fishmeal is an expensive practical ingredient and cost usually limits the level of incorporation in the diet. Given that feeds are the single largest variable cost in most pond aquaculture operations and that fishmeal is the most expensive component of aquaculture feeds, there is a strong economic incentive to reduce feed costs and reduce the level of fishmeal in feeds formulated with practical ingredients. Fishmeal replacement remains one of the most active areas of aquaculture nutrition research.

Environmental advocates have criticized the use of fishmeal in aquaculture (Naylor et al. 2000), claiming that the production of fish in aquaculture using fishmeal derived from capture fisheries for small pelagic species represents a net loss of protein. These arguments have been refuted on the basis of the differences in the ecological efficiency of carnivorous fish growth in natural and aquaculture production environments, and on the greater efficiency of protein production in aquaculture compared to terrestrial livestock production, where feeds include fishmeal. However, the use of fishmeal in aquaculture feeds remains controversial. Other studies (Hites et al. 2004) have indicated that levels of environmental contaminants in fishmeal are variable, but they can accumulate and concentrate in cultured fish. Thus, reducing the level of fishmeal in aquaculture feeds, particularly from sources with high levels of environmental contaminants, is desirable from the standpoint of food safety and product marketing.

Meals made from the rendered by-products of beef or pork processing are nutritionally good substitutes for fishmeal in feeds for most pond-raised fish. Although animal protein sources other than fishmeal are good sources of dietary energy, protein, and minerals for fish, some consumers may be concerned about fish grown on feeds containing by-products of ruminant livestock slaughter because of a perceived link with "mad-cow" disease (bovine spongiform encephalopathy or BSE). In response to consumer concerns, the European Union Commission on Food Safety and Animal Welfare strengthened the European Union's BSE control measures in June 2000 by requiring all member states to remove ruminant animal by-products from all animal feeds, including fish feeds. The United States Food and Drug Administration bans the use of most rendered mammalian proteins in feeds for ruminant animals but allows their use in feeds for nonruminants, including poultry, swine, and fish because there is no evidence of transmission of BSE from ruminants to nonruminant animals (Matthews and Cooke 2003; Ingrosso et al. 2006).

In summary, the reasons for reducing the amount of fishmeal and animal protein in aquaculture feeds are related to 1) incentives to reduce the cost of production, 2) negative consumer perceptions about the incorporation of animal protein in general and fishmeal in particular into aquaculture feeds, and 3) perceived or real environmental impacts on natural ecosystems of which small pelagic fish species are a key component.

Feeds for omnivorous species such as channel catfish and herbivorous species such as tilapia do not require animal protein to meet dietary amino acid requirements. Nonetheless, some fishmeal (up to 4%) and fish oil (as a top dressing) is included to increase the palatability of the feed. Feeds for omnivorous and herbivorous species should not include animal protein from the by-products of the slaughter of ruminant livestock.

Handle and store feeds to maintain feed quality

Feeds for commercial aquaculture are formulated and

manufactured to have low amounts of inedible fi nes and to remain stable until they are used. Feeds must, however, be handled and stored properly to prevent deterioration and to maximize the efficiency of feed use by the cultured species. Bagged feed should never be handled roughly because abrasion of feed within the bags can produce large amounts of fi nes that will not be consumed by fish and therefore will add directly to the waste load within the pond. Bulk and bagged feeds should be stored in cool, dry areas. Because this may not be pos-sible with bulk feed stored in bins on the farm, bulk feed should be purchased in amounts that will be used quickly so that storage time in minimized. Use feeds on a fi rst-in and fi rst-out basis so that older feed in bins or bags is used before using more recently pur-chased feed. Improperly stored feed or feed stored for too long may become moldy. Never feed moldy feed to fish because molds reduce the nutritional quality of the feed and may produce toxins detrimental to fish growth and health. Barrows and Hardy (2001) present an excellent detailed account of proper feed handling and storage, particularly for bagged feeds.

POND FERTILIZATION

Chemical or organic fertilizers are applied to some ponds to increase aquaculture produc-tion. Chemical fertilizers provide essential plant nutrients that stimulate phytoplankton growth; organic fertilizers provide nutrients and organic matter that stimulates plant and bacteria growth. Phytoplankton and bacteria then serve as the base of food webs that provide natural food organisms—such as zooplankton and insects—for fish or crustaceans. Pond fertilization is common throughout the world in the culture of fish and crustaceans that are capable of effi ciently using natural food production. In the United States, pond fertilization is most commonly used in sportfish ponds and aquaculture ponds used for bait and ornamental fish culture. Combinations of organic and chemical fertilizers are also used in ponds for the early life stages of hybrid striped bass, channel catfish, and other foodfish. Fertilization is also used to promote phytoplankton blooms that prevent the growth of nuisance weeds and to stimulate production of a vegetative forage crop that serves as the base of a detrital food web in commercial crayfish ponds.

When used correctly, fertilization can be an ecologically effi cient, cost-effective approach to pond aquaculture. Misuse of fertilizers can, however, be wasteful of a valuable resource and increase the risk of environmental degradation inside and outside the pond.

Better Management Practices for Pond Fertilization

When used to promote phytoplankton blooms, apply fertilizers only as needed

Ponds receiving large inputs of nutrients from other sources should not be fertilized. For example, plant nutrients derived from fish wastes via manufactured feed will usually promote abundant phytoplankton growth without fertilization. Likewise, ponds with water-sheds in pastures of agricultural cultivation may receive nutrient-enriched runoff that encourages adequate phytoplankton growth without additional nutrient input. In either case, fertilization may cause excessive phytoplankton growth; greater oxygen demand; and higher concentrations of organic matter, nitrogen, and phosphorus in effl uents. An exception is the use of fertilizers to encourage rapid phytoplankton growth for aquatic weed control immediately after ponds are filled. Fertilization for aquatic weed control in pasture ponds or ponds with feeding can be stopped once a phytoplankton bloom develops. The bloom will then be sustained by nutrients derived from feed or runoff.

Use fertilizers effi ciently

Ineffi cient use of fertilizers is wasteful, economically unsound, and may result in excessive phytoplankton growth and elevated nutrient and solids concentrations in pond effl uent. Regrettably, effi cient pond fertilization practices are difficult to determine because response to fertilizers varies depending on water source (particularly hardness and alkalinity), texture and chemistry of pond bottom soil, watershed land use, pond morphology and hydrology, climate, past fertilization history, and a host of other factors. Fertilization recipes are available for ponds in some regions but these recommendations are, at best, only guidelines that may need to be changed depending on the response in a particular setting (Box 6.9). Boyd and Tucker (1998) summarize pond fertilization programs and also discuss factors affecting pond fertilization success. Private consultants or experts at universities can provide advice on fertilization practices applicable to specific regions and culture species. Basic soil and water testing can indicate the need for specific nutrients and appropriate types, rates, and frequency of fertilizer application.

Apply agricultural limestone to ponds with total alkalinity below 20 mg/L as $CaCO_3$

Acidic, low-alkalinity water is perhaps the most common cause of poor response to pond fertilization and ineffi cient fertilizer use. Acidic bottom soils adsorb fertilizer phosphorus, making it unavailable to phytoplankton. Low-alkalinity waters are also defi cient in carbon, an essential plant nutrient. As a general rule, fertilizers are not effective in promoting phytoplankton growth in pond waters with total alkalinities below 20 mg/L as $CaCO_3$. The problem is easily corrected by applying agricultural limestone to increase alkalinity (Boyd and Tucker 1998).

Do not use animal manures for fertilizers

Organic pond fertilizers include composted vegetation, hay, grain and oilseed meals, or animal manures. Organic materials added to ponds slowly decay and release nutrients that are then available to support phytoplankton growth. Organic detritus produced during decomposition may also be consumed by zooplankton, macroinvertebrates, and even by the cultured fish or shellfish. Organic fertilizers often make use of low-quality organic material that would otherwise be considered a waste product. Combined with chemical fertilizers, organic fertilizers promote a stable and diverse plankton com-munity that is particularly desirable in the culture of larval fish and juvenile fish (Ludwig 2004).

Organic fertilizers are not widely used in the United States, the exception being grain and oilseed meals used as part of fertilization programs for the early life stages of some foodfish. Most organic materials have low fertilizer value and are less convenient to use than chemical fertilizers. Organic fertilizers also create an oxygen demand as they decom-pose and release humic substances during decomposition that may stain the water.

When organic fertilizers are needed, grains or oilseed meals are preferred to manures or other wastes. Animal manures may contain antibiotics or other drugs that could con-taminate effl uents or fish fl esh. Animal manures are also nutrient-poor organic fertilizers and may cause oxygen depletion when used in amounts necessary to promote rapid phy-toplankton growth. Other aesthetic and sanitary concerns related to the use of manures argue against their use in foodfish aquaculture, despite their widespread use in low-input aquaculture throughout the world.

Do not exchange water after fertilization

Water exchange reduces the effectiveness of fertilization by fl ushing and diluting nutrients and natural food organisms. Water exchange also increases mass discharge of organic matter and nutrients by increasing effl uent volume. Fertilization rates, frequency of fertil-izer application, and mass discharge of nutrients and organic matter can therefore be reduced by minimizing pond overflow. Practices for reducing pond overflow are discussed in the section "Overflow Effl uents."

Do not fertilize ponds 1 or 2 days before significant precipitation events

Concentrations of plant nutrients are highest a day or two after chemical fertilizers are applied to ponds. If pond overflow occurs at that time, the nutrient concentration in effl u-ent will be greater than normal. Runoff following heavy rains can also muddy ponds and impair response to fertilization because of inadequate light for phytoplankton growth. Clay particles in turbid runoff sequester phosphorus and remove the nutrient from water as clay particles settle to the bottom or are washed out of the pond. Although it is difficult to predict short-term weather, attention should be paid to offi cial forecasts and pond fertiliza-tion should be postponed if there is a significant chance that heavy rainfall will cause overflow or muddy conditions in the pond.

Store fertilizers properly

Fertilizer storage areas should be separate from other uses. Store fertilizers on pallets in a cool, dry place. The storage area must be covered and protected from rain and runoff. Preferably, the storage area should be a building with an impermeable fl oor. Store fertil-izers away from pesticides or other agricultural chemicals and purchase fertilizers in amounts that will be used relatively quickly so there is no carryover from year to year. Spills or ripped bags should be immediately swept up and stored in barrels or buckets and used fi rst during the next scheduled application.

FISH ESCAPE

Fish can escape when aquaculture ponds become hydrologically linked to adjacent receiv-ing waters. In most aquaculture ponds under normal operating conditions this connection is intermittent because water is discharged only when ponds are drained or overflow after heavy rains. However, hydrological connection can occur as a result of design, construc-tion, or operational failures. Ponds may be improperly sited in fl ood-prone locations, where fish may escape when fl oodwaters overtop pond embankments that were not constructed sufficiently high. If ponds are not properly constructed, embankments or drainage struc-tures may fail, leading to the release of fish. Fish can also escape during routine farm activities, especially when fish are handled during stocking, grading, or harvesting. Fish can also escape through predation or vandalism.

The cumulative risks of escape must be considered in terms of the hazards or conse-quences associated with escape and the probability of each hazard occurring. The main hazards associated with escape include displacement of native species through direct pre-dation, competition for food or spawning sites, or habitat alteration; transmission of pathogens to natural populations; and potential changes in allele frequencies within popu-lations of native conspecifics. In the case of channel catfish culture in ponds in most of North America, the geographic extent of pond aquaculture overlaps with the native range of the species, and so the consequences of escape to natural populations are relatively benign. In the case of ornamental fish culture in Florida, virtually all species cultured are not native, and at least 20 of the 32 nonnative fish species that have been released or escaped are now established in that state.

Nonnative fish that escape, persist, and cause environmental and economic harm when they escape are described as invasive. Invasive species have characteristics or a combination of characteristics that facilitate their survival and establishment: broad water quality tolerance limits, omnivory, rapid growth rates, early maturity, high fecundity, and a lack of natural competitors or predators. Invasive species can have negative effects on biodiversity, habitat quality, and ecosystem functioning. Regrettably, the record of species that have escaped from pond aquaculture facilities and established reproducing populations is not good. There are notable examples of the establishment of invasive species that were introduced to a watershed by escape from aquaculture ponds. In the United States, the establishment of grass carp, bighead carp, and silver carp in the Mississippi River and its main tributaries is attributed, in part, to escapes from commercial pond aquaculture facilities. A variety of species have been purposefully stocked as gamefish by public agencies that may also have escaped commercial culture, such as common carp; brown and rainbow trout; largemouth bass; and various sunfishes, pikes, salmonids, and catfish. Escaped fish represent an undesirable loss of potential revenue to pond culturists, so there is a strong economic incentive to prevent the occurrence of escapes. Thus, the practices discussed below focus on measures that can be implemented to prevent escapes and not measures to contain the spread of fish once escape has occurred.

Better Management Practices for Preventing Escapes

Consider the risk of fl ooding in site selection

Selecting sites that are subject to fl ooding increases the risk of escape if the elevation of fl oodwaters exceeds that of the top of pond embankments. Sites should be selected above the 100-year fl ood elevation as determined by the USDA Natural Resources Conservation Service or an equivalent natural resource regulatory agency. The elevation of the top of pond embankments should be 50 cm above the 100-year fl ood elevation. Selected sites may experience 0 to 5 fl oods every 100 years.

Construct and maintain pond embankments to prevent failure

Standards for pond design and construction as outlined in the pond construction section should be followed. Pond embankments should be constructed with 3 : 1 side slopes to minimize erosion and scouring by fl oodwaters. Embankments should be compacted prop-erly. Embankments that are subject to erosion from wind-driven waves should be armored with riprap or other protective measures. Embankments should be vegetated shortly after construction and grass cover maintained by mowing. Burrowing animals should be con-trolled by trapping or hunting. Documented levee inspections should occur at least monthly and following each severe storm event.

Include barriers to escaped fish in drainage structures and ditches

Barriers to the escape of fish should be considered optional for species that are cultured within their native range, but essential for nonnative species. No single barrier will prove sufficient and a redundancy of barriers should be built and maintained to ensure effective-ness. Barriers must be sufficiently open to allow water to flow through, but sufficiently closed to retain the smallest life stage (e.g., egg, larvae, fry, or fi ngerling) in the pond. Debris must be regularly cleared from all barriers and screens to prevent clogging and barrier overtopping during overflow.

Pipe culvert barriers can be placed over inlet or outlet ends of drainpipes or the pipe risers that control water level. For fry, 5-mm (3/16-inch) mesh, and for fi ngerlings, 13-mm (1/2-inch) mesh can be secured to pipes with a coupling band or clamp. This barrier is appropriate for flows to 1.7 m^3/minute (1 cubic foot per second).

Spillway barriers consist of a metal framework with horizontal bars or expanded metal mesh extending the full width of the spillway and 1 m above the normal high-water level. The barrier should be constructed to withstand a 20-year fl ood. The horizontal bars can be 9.5- to 25-mm (3/8- to 1-inch) diameter metal bars or PVC pipe. To retain grass carp, bars should be spaced 25 to 38 mm (1 to 1.5 inches) apart. Gabion fi lters (rock screen) consist of various-size rocks (10 to 20 cm; 4 to 8 inches) and gravel encased in wire mesh or chain-link fencing that is placed on the spillway. Gabion fi lters that are 2 m wide × 3 m long can accommodate water flow rates less than 1.7 m^3/minute (1 cubic foot per second).

Drainage ditch screens consist of fl attened, expanded metal mesh welded to an angle-iron frame that extends across the full width of a drainage ditch. The barrier should be placed fl ush against ditch side slopes. Barrier height should be sufficient to handle inter-mittent flows to 3.4 m^3/min (2 cubic feet per second). If on-farm ditches or stormwater retention ponds hold water year-round, they should be stocked with native predaceous fish that will consume escaped fish.

Prevent escape during fish transfers

Production of most species in pond aquaculture is accomplished in phases, necessitating movement of fish between one culture unit and another. For example, channel catfish fry are stocked into nursery ponds from a hatchery and then harvested and stocked into food-fish production ponds. At harvest, foodfish are graded, loaded into transport trucks, and hauled to a processing plant. At each point where fish are handled, there is opportunity for escape. In particular, fish can escape with outflowing water when ponds are drained for harvest if drainage structures are inadequately screened. Knowledge of the behavior of fish in response to flowing water can help determine what type of barrier is appropriate. Fish escape during transport can be avoided by maintaining harvest and transport equip-ment in good working order through periodic inspections and restricting operation of loading equipment to trained personnel. An HAACP-type evaluation can assess the steps in the production process where the probability of escape is elevated, leading to imple-mentation of a preventive program to minimize that probability.

Deter nuisance wildlife

Active nuisance wildlife deterrence is an essential component of an escape prevention program. Burrowing animals, such as muskrat, nutria, and beaver, can excavate tunnels in embankments, increasing the risk of washout, and undermining the structural integrity of the embankment. Emergent aquatic plants, especially those with starchy tubers, that provide food and cover should be eliminated from pond margins. Keeping embankments well mowed or periodic spot applications of approved herbicides can control emergent aquatic plants. Burrowing animals can be captured with traps set in den openings, runways, or slides. After animals are removed, the area of the embankment around the burrow opening should be excavated to the full extent of the burrow, refilled with soil, properly compacted, and promptly revegetated to minimize erosion.

Although not as serious a potential problem as escapes caused by burrowing animals, some birds can carry live fish from ponds. Most of the bird species that are most problematic from the standpoint of fish predation consume fish at the point of capture. However, some raptors (e.g., ospreys and eagles) and owls capture fish swimming near the pond surface by grasping and holding with talons and fl ying off to a perch to consume the fish. Struggling fish may be dropped into receiving waters outside the facility bound-aries. Although a limited number of some species of birds can be killed under permits, raptors are protected by legislation, and so conventional, nonlethal scare tactics are the only option. Effective nonlethal methods include frightening devices (e.g., propane cannons, pyrotechnics) in combination with harassment patrols. Nuisance birds can be excluded from small ponds by overhead wires, nets, or other barriers. In the United States, the USDA Animal and Plant Health Inspection Service (APHIS), Wildlife Services program provides technical assistance on appropriate control methods for nuisance wildlife.

Do not culture invasive species in ponds without rigid safeguards

Pond aquaculture of potentially invasive species should only be undertaken after thorough study and consultation with knowledgeable fisheries and aquaculture specialists and federal and state regulators to determine risks and benefits. It is rare that the potential benefits of culturing an invasive species will outweigh the costs associated with escape and invasion. Due diligence should be documented when deciding whether culture of an invasive species is appropriate. Producers should demonstrate that adequate protection measures are in place and that the risk of escape and establishment is low. Producers should review the fish life history, habitat and food preferences, and the history of invasion elsewhere of a particular species to assess the risk of establishment.

State fish and wildlife resource management agencies usually administer permits to culture nonnative species. The United States Fish and Wildlife Service can designate invasive species that have become established and caused ecological and economic harm as injurious, which limits their importation into the United States and prohibits interstate transport.

Beyond physical barriers, producers of nonnative species should make a conscientious effort to prevent escape and mitigate postescape effects of nonnative species. For some species, sterile (triploid) fish or monosex fish can be cultured. A plan should be in place in the event of invasive species escape to contain, recover, or eradicate them. Recapture equipment or sufficient quantities of approved piscicides should be in place for a rapid response, if this approach is warranted and allowed by law.

PREDATOR CONTROL

High concentrations of fish and crustaceans in outdoor aquaculture facilities provide attrac-tive foraging opportunities for certain species of birds, reptiles, mammals, and fish. Relative to other aquaculture production systems, ponds are especially vulnerable to wildlife dep-redation. Ponds may extend over large areas that are difficult or impossible to protect using barriers or enclosures. Pond aquaculture facilities are usually located in rural or isolated areas that already support diverse and abundant wildlife, and many of these animals will be attracted to the concentrated food resources available in ponds. Ponds constructed in the fl yways for migratory birds offer convenient foraging opportunities and, in areas with extensive aquaculture development, the attraction of a year-round food supply may even cause birds to change foraging behavior and migration patterns and linger near farms for longer periods of time (Glahn and King 2004).

The effect of predators on aquaculture production can range from insignificant to cata-strophic. Mammals (such as otters [*Lontra canadensis*], mink [*Mustela vision*], and nutria [*Myocaster coypu*]) and reptiles (such as snakes and alligators) may be a nuisance, but they rarely cause significant economic losses in pond aquaculture.

Predacious fish can cause serious losses, especially in the early phases of production. Birds are the most impor-tant predators in pond aquaculture, although impact varies with species and number of birds. Certain birds, such as the great egret (*Ardea alba*), feed primarily on weak or dis-eased fish found near pond banks and their economic impact is probably minor. In contrast, double-crested cormorants (*Phalacrocorax auritus*) are highly adapted to foraging on live fish in open water and often visit aquaculture ponds in large fl ocks. Cormorant depredation may therefore cause serious economic hardship in areas with abundant populations.

In addition to direct losses to predation, predators indirectly affect aquaculture produc-tion by serving as vectors for infectious diseases. Birds and other animals may move infected fish from one pond to another or spread pathogens in regurgitated stomach con-tents or fecal material. The importance of animals as mechanical vectors of disease is, however, largely unknown. On the other hand, birds also serve as obligate hosts for certain digenetic trematode fish parasites and are therefore integral in the spread of these disease organisms among water bodies. Digenetic trematodes have complex life cycles in which fish may act as either intermediate hosts that harbor an immature stage of the parasite or as fi nal hosts in which the adult trematodes are found. Trematodes using fish as interme-diate hosts usually involve fish-eating birds as fi nal hosts and mollusks, usually snails, as fi rst intermediate hosts.

Predator control practices should minimize the interaction between wildlife and cultured animals to protect wildlife and minimize production losses. Predators may not only cause direct loss of fish, but the mere presence of predators, especially birds, around aquaculture facilities can keep fish in a perpetual state of fright, resulting in increased physiological stress and poor feed conversion. Direct and indirect effects on production will reduce resource-use efficiency and increase waste generation. Limiting access of predators to moribund and dead fish minimizes the potential spread of pathogens into wild fish populations.

Given the abundance of wildlife, especially birds, associated with aquaculture ponds, it is highly unlikely that predator control practices in pond aquaculture will negatively impact natural populations of wildlife. In fact, the conversion of agricultural land to aqua- culture ponds has created additional habitat for wading birds, waterfowl, and other species that has helped support healthy populations in the face of wetland and other habitat loss elsewhere (Huner 2000). Nonetheless, aquaculture ponds are visited by birds that are clas-sifi ed as threatened or endangered by the United States Fish and Wildlife Service, and aquaculturists must be aware of the legal and ethical limitations on predator control methods for threatened and endangered species.

Predation of cultured fish or crustaceans can be eliminated only through total exclusion, which is impractical on all but the smallest pond facilities. Also, many predators quickly learn to avoid or ignore management efforts. Effective management usually relies on a combination of management approaches. Overall, the best approach is one based on prin-ciples of Integrated Pest Management where predators are identifi ed, the type and level of damage is assessed through regular monitoring, and control methods are chosen that are appropriate for the predator and level of damage.

Better Management Practices for Predator Control

Assess predator impact

The potential impact of predation determines whether management efforts are economi-cally justifi ed. Predatory mammals tend to be nocturnal and elusive, and often the only signs of predation are tracks, scat, and chewed or partially eaten fish on pond embank-ments. Although birds often swallow fish whole, leaving few signs of damage, predation by birds is easy to confi rm because the important fish-eating birds, except for pelicans, forage during daylight hours. Problems can be initially confi rmed by observing actively feeding birds or birds loafi ng or roosting near ponds.

Regrettably there are no simple quantitative tools to assess the economic impact of wildlife damage on aquaculture farms, and there is no way to predict whether a small problem will become a major problem if action is not taken. It is therefore prudent to seek advice and assistance from experts as soon as a problem is suspected. Technical assistance and information on wildlife predation control at aquaculture facilities can be obtained by contacting the nearest USDA Animal and Plant Health Inspection Service, Wildlife Services offi ce. To fi nd the nearest Wildlife Services offi ce, contact your local county Extension agent or call the USDA-APHIS/WS Operational Support Staff Offi ce in Riverdale, Maryland, at (301) 734-7921.

Problems with digenetic trematodes are a special case because widespread infections of fish may occur with little or no evidence of bird involvement. Birds that serve as fi nal hosts for adult trematodes need not feed on fish in a pond to initiate an infection. If a single trematode-infected bird fl ies over a pond and defecates, thousand of trematode eggs will be dispersed into the pond. If the pond contains the correct snail and fish intermediate hosts, the life cycle can be completed and fish will become infected. Moreover, American white pelicans (*Pelecanus erythrorhynchos*)—the fi nal host for the *Bolbophorus damnifi -cus* trematode that infects channel catfish—often forage at night and farmers may be unaware that pelicans are visiting or fl ying over their farm. Risks of trematode infections are best assessed by frequent monitoring of the fish population for clinical signs of infec-tion. Farmers should also be aware of early, population-level signs of the disease, such as an unexplained reduction in fish feeding response (Wise et al. 2004).

Identify the predator responsible for losses

Identifi cation of the species responsible for predation is necessary to assess the potential impact of predation and formulate a management plan (Fig. 6.9). Species identifi cation is also critical because wildlife management is regulated by federal and state laws. Field guides to birds and other wildlife are available at local libraries and bookstores. Technical assistance can be obtained from USDA Wildlife Services using the contact information above.

Check with appropriate regulatory authorities

All fish-eating birds are protected under the Migratory Bird Treaty Act and may not be killed without a depredation permit or depredation order. The United States Fish and Wildlife Service (USFWS) has regulatory authority for managing migratory birds. If fish-eating birds are causing problems at an aquaculture facility, the USFWS may issue a depredation permit that allows the producer to kill a limited number of some species to reinforce the effectiveness of nonlethal control measures. Some fish-eating birds also are protected by the Endangered Species Act.

For example, wood storks (*Mycteria americana*) found east of the Mississippi-Alabama border receive this protection. No lethal or nonlethal control activities can be used to control any bird species using aquaculture facilities in this region if wood storks are nearby. The Bald and Golden Eagle Protection Act further protects eagles and prohibits all hazing activities near bald (*Haliaeetus leucocephalus*) and golden eagles (*Aquila chrysaetos*), except with special permission from the USFWS.

Federal and state laws may also pertain to other wildlife, and the fi rst step when con-sidering managing predation problems on aquaculture facilities should be to contact USDA Wildlife Services using the contact information above. Operational support staff at Wildlife Services can assist with identifying the problem and recommending legal control methods.

Design facilities to reduce predation losses

Physical characteristics of farms can affect the extent of predation losses and the effective-ness of predator-control practices. Isolated farms and farms near loafi ng or roost sites are especially vulnerable to bird predation. Small fish are more susceptible to wildlife predation than larger fish, and on farms with both fi ngerling and foodfish ponds, fi ngerling ponds can be located near the farm headquarters or other areas with more human activity. Birds and other wildlife tend to avoid close proximity to humans, and ponds closer to headquar-ters will also be more convenient to patrol and manage for wildlife control. Large ponds tend to suffer greater bird predation losses because birds feel safer on large ponds and management is more difficult. For example, birds harassed from one embankment or one end of a large pond often simply fl y across the pond and resume feeding. Using smaller ponds—particularly for small, vulnerable fish—can reduce losses to wildlife.

It may not be practical to redesign or relocate existing facilities to address predation problems, but even relatively minor changes can help reduce losses or make management efforts more effective. For example, removing weed and brush around ponds, keeping embankments mowed, and selectively removing nearby trees that serve as loafi ng, hiding, or roost sites can reduce predation pressures. Assuring all-weather access to ponds is also important so that harassment efforts can continue without interruption regardless of weather.

Use frightening or harassment techniques

Frightening and harassment techniques use sound, sight, or both to discourage birds and other wildlife from feeding, roosting, or loafi ng near ponds. Passive frightening and harass-ment programs (those not requiring human presence to function) employ a wide variety of devices, including exploding or whistling pyrotechnics, automatic propane cannons, fl ashing lights, human effi gies (scarecrows), fl ash tape, and even old vehicles parked on pond levees. Programs should be initiated before birds arrive and establish feeding habits. For example, on southern aquaculture farms visited every year by overwintering double-crested cormorants, deterrence programs should start in fall when birds fi rst begin arriving from northern breeding grounds.

Harassment must be aggressive and persistent to be effective. Most passive techniques discourage predation only for a short time because birds quickly lose their initial fear and become habituated to the deterrent device. Periodically changing the location of the devices or alternating the type of device can enhance long-term effectiveness. However, even the most aggressive passive deterrence program rarely eliminates wildlife predation and may quickly become ineffective in areas with heavy predation pressure. Active patrols employing nonlethal or lethal measures to reinforce passive measures are usually required for effective, long-term control, especially on large farms.

When all measures to disperse birds using nonlethal techniques have been exhausted, farmers may consider using limited killing of birds to reinforce the fear of nonlethal mea-sures. Depredation permits are required from the United States Fish and Wildlife Service and, in some states, from the state wildlife agency to kill almost any species of bird. For currently applicable laws, contact the nearest USDA Wildlife Services or United States Fish and Wildlife offi ce.

Discourage birds from areas near the farm

Fish-eating birds usually focus daily activities around night roosts or daytime loafi ng sites within a short fl ight to feeding areas. Dispersing birds from roosts or resting sites can augment on-farm deterrence activities, although the value of off-farm dispersal varies with species and intensity of the management effort. Cormorants, for example, usually have night roosts within a few miles of foraging areas, and regionally coordinated roost dispersal activities can significantly reduce numbers of cormorants visiting catfish farms (Tobin et al. 2002). Pelicans may forage near their daytime loafi ng areas, but foraging activities range over a much greater area than cormorants and are more difficult to manage.

Dispersing cormorants from roosting and loafi ng areas may require teams working in a regionally coordinated effort. Teams should enter the roost shortly before sunset and fi re pyrotechnics into the roost to prevent birds from settling. Roosts should be dispersed on several successive nights. If there are several roosts in the immediate vicinity of a farm where damage is occurring, all the roosts should be dispersed simultaneously so that birds do not simply fl y to another nearby roost.

American white pelicans use shallow water areas such as rice fields, waterfowl impound-ments, fl ooded fields, and abandoned catfish ponds as daytime loafi ng areas. Loafi ng areas can be made less attractive by patrolling daily and harassing birds to ensure that they are dispersed. Abandoned fish ponds should be drained to eliminate them as habitat for fish-eating birds.

Use exclusions, impediments, or barriers

Total exclusion is the only means of eliminating predatory wildlife and may be justifi ed on small farms raising valuable crops, such as ornamental fish. Exclusion structures are usually made of netting, which should be a dark material rather than clear monofi lament to reduce chances that birds will inadvertently fl y into the netting and become entangled. Netting should also be checked frequently to remove any trapped birds. Exclusion struc-tures should be installed by knowledgeable professionals to assure ease of maintenance and resistance to damage by ice and wind. Relatively inexpensive polypropylene or nylon netting that is specifically manufactured for predator exclusion from aquaculture facilities can be installed (Fig. 6.10). Depending on the size of the facility and the durability of the structure, netting enclosures may require a relatively large initial investment, but the cost can be amortized over a relatively long period.

Complete exclusion of predators is not possible on ponds larger than 1 to 2 ha and is not economically feasible on large farms, regardless of pond size. Partial exclusion or deterrence using impediments such as wires, lines, or fencing may be more cost-effective on larger facilities. Such devices do not prevent bird predation, but rather reduce predation by discouraging birds from landing or wading into ponds. Overhead lines and wires suspended in one direction or in a crossing pattern are particularly effective in reducing depredations by birds with long takeoff distances, such as cormorants. Lines can be widely spaced (10 to 20 m) and should be highly visible to the birds to be effective. Adding brightly colored streamers or Mylar balloons increases the effectiveness of these systems. Major constraints to using such impediments are cost and interference with pond culture practices, especially fish harvesting.

Wading birds, such as herons and egrets, prefer to land on solid ground and walk into the pond. Fencing or perimeter netting will discourage wading birds from walking into the pond. Perimeter fences may not be effective if the pond margins are shallow because birds can land and forage on the water side of the barrier.

Exclusion devices are expensive and there are numerous options. Before making an investment in a particular technique, farmers should obtain technical assistance from the nearest USDA Wildlife Services offi ce using the contact information above.

Prevent the introduction of predatory fish

Predatory fish accidentally introduced into aquaculture ponds can cause large losses, especially in ponds used to raise bait or ornamental fish or in nursery ponds for juvenile catfish or hybrid striped bass. Aside from direct losses attributable to predation, wild fish can also serve as vectors for infectious disease organisms. The key to managing losses to predacious fish is to prevent their introduction, because it is difficult and expensive to selectively remove wild fish from culture ponds using piscicidal chemicals.

Unwanted fish most commonly enter ponds with the water supply, as contaminants during stocking, or through unsecured pond drainage structures. Additional routes include fl oodwaters that overtop pond embankments and fish transported from one water body to another by wildlife. Many of the practices applicable to controlling accidental escape of the culture species are also applicable to controlling accidental introduction of predacious fish. Ponds should be properly constructed in areas not prone to fl ooding, drainage struc-tures should include barriers to prevent fish movement into or out of the pond, and nuisance wildlife should be deterred. The risk of accidentally introducing wild fish can be reduced by using groundwater to fill ponds. If surface waters are used, inlets should be screened to prevent introduction of fish or eggs. Any undesirable fish in the pond can be eliminated by draining and drying the pond before refilling for the next culture cycle. Fish in puddles remaining when the pond is drained can be eliminated by applications of quicklime. Fry and fi ngerlings should be purchased from reputable sources and inspected for contamina-tion with unwanted fish before stocking.

AQUATIC PLANT CONTROL

Aquatic plants are essential components of pond aquaculture systems. In cultures relying upon natural food organisms to support aquaculture production, plants are the base of the food web and their growth is often encouraged through fertilization to increase aquaculture yield. Plant communities are also important in pond cultures receiving manufactured feed, even though plant-based food webs are of minor importance to the nutrition of animals in those cultures. Some plant growth is necessary because plants function with other com-ponents of the pond ecosystem to help maintain adequate environmental quality for aqua-culture production. For example, oxygen produced in plant photosynthesis is usually the major source of dissolved oxygen in ponds.

Despite the importance of plants in aquaculture ponds, some plant communities interfere with pond management (especially harvesting and feeding), endanger the well-being of the animal under culture, or impair the quality of the aquaculture product. Such communi-ties are undesirable, and aquaculturists often attempt to prevent them from developing in ponds or try to eradicate them if they become established. Aquatic plant management is a common practice in pond aquaculture and involves a variety of practices, all of which have potential environmental impacts both in and outside the pond. The overall goal of plant management should be to control undesirable vegetation in a way that is compatible with maintaining a well-functioning environment for fish production with minimal or no effects on nontarget organisms inside and outside the pond. Effective, environmentally sound management of aquatic plant communities can be difficult because the ecology of aquatic plants is complex. Effective management, espe-cially when using herbicides, requires experience and training. Consult with aquatic plant control experts at local universities or with lake-management consultants before undertak-ing treatment. Plant identifi cation keys and management guidelines are available in Westerdahl and Getsinger (1988) and Petty (2005). Excellent online resources for aquatic plant management include websites managed by Texas A&M University's Aquaplant (aquaplant.tamu.edu) and the University of Florida's Center for Aquatic and Invasive Plants (aquat1.ifas.ufl .edu).

Better Management Practices for Aquatic Plant Control

Prevent weed problems whenever possible

Almost any plant can be tolerated as long as it does not become so abundant that it inter-feres with producing the aquaculture crop. It is, however, difficult to predict whether a small infestation of weeds will spread and become a problem, so most control measures are implemented only when fairly large stands of weeds have already become established. At that time, using chemical herbicides is usually the fastest way to eradicate weeds and reestablish a phytoplankton community, which is usually the most desirable plant form in aquaculture ponds. Chemical weed control is, however, risky in aquaculture ponds because water quality deteriorates when dense stands of weeds are killed. Certain practices can reduce the probability of weed problems, thereby reducing chemical use.

Most noxious weeds start growing in the shallow areas of ponds where light penetrates to the bottom. Deepening the edges of ponds will discourage the establishment of emergent or submersed weeds, although this may not be possible in large ponds subject to rapid pond bank erosion. Ponds should have an adequate supply of water and be filled as quickly as possible. If several ponds are served by a common water supply, it is better to fill them one at a time rather than to slowly fill several ponds at once. By increasing water depth as quickly as possible, plants that grow from the pond bottom have less opportunity to become established. Excessive water flow through ponds fl ushes plant nutrients from the water, favoring rooted weeds that can obtain nutrients from bottom soils. Ponds should not have watershed areas larger than necessary to maintain water level and excess runoff from large watersheds should be diverted away from ponds.

An effective fertilization program can encourage the development of a phytoplankton bloom. A well-established phytoplankton community competes with other types of plants for light and nutrients, and usually prevents growth of submersed and emergent plants in all but the most shallow water. Fertilization should only be used to prevent weed infestá-tions; if noxious stands of weeds are already established, fertilization will make conditions worse by stimulating further weed growth.

Removing weeds mechanically or by hand may reduce the possibility of having to use other measures to control plants. Routine mowing of pond banks will help prevent the establishment of dense growths of shoreline plants such as willows and will also reduce habitat for snakes. Manual removal of plants is feasible only in small ponds. Plants should be removed immediately as small areas become infested, because it is time consuming and laborious to manually remove established stands of weeds. Care should be taken to remove as much of the rootstock or rhizome as possible to minimize regrowth. Preferably, manual removal should be undertaken in the beginning of the plant-growing season.

Identify the weed problem

Correct identifi cation of weeds is critical because potential impacts and management differ for each plant. Strategies effective on one species may be ineffective even on similar species. In particular, herbicides are selective (some much more than others), and effective control depends on matching the weed with the most appropriate herbicide. Private con-sultants or experts at local universities can help identify the weed problem. Plant identi-fi cation keys are available in the publications and websites listed above.

Make management decisions based on site-specific conditions

Environmentally sound and cost-effective management depends on the type of plant, the extent of plant coverage, the species and life stage of fish or crustaceans in the pond, water quality, time of year, and weather. Understanding these interactions, which differ for each weed problem, is largely a matter of experience. Until that experience is gained, seek advice from private consultants or experts at universities.

Use only those herbicides labeled for use in aquaculture

Although hundreds of chemicals are used for weed control in the United States, only about ten active ingredients are labeled for use in aquaculture. Herbicide trade names and the status of herbicide label clearances change over time, so be sure to review the most current herbicide information for proper use and restrictions.

Carefully follow herbicide labels

Herbicides sold in the United States must be registered with federal and state regulatory agencies. The printed information accompanying the herbicide container is called the *label* and constitutes a legal document. Failure to use herbicides according to label instructions can lead to severe penalties. From a practical standpoint, misuse of herbicides can result in poor weed control; risks to people, the aquaculture crop, or wildlife; or herbicide residue problems in the aquaculture product.

The label provides information on the active ingredient, directions for correct use on target plant species, warnings and use restrictions, and safety and antidote information. Remember, state and local regulations may be more restrictive than federal regulations. Certain products are registered as Restricted Use herbicides and can be legally applied only by trained and certifi ed applicators or by people under their direct supervision. Be sure to check federal, state, and local regulations prior to using herbicides.

Herbicide treatment rates are based on pond area or pond volume. Miscalculation will result in either overtreatment or undertreatment (which may require additional treatments to eradicate the weed). In either case, more chemical than needed will be applied to the pond. Carefully measure pond dimensions and keep up-to-date records of pond size and depth. Pond surface area tends to increase and pond depth tends to decrease over time because of erosion of embankments and sedimentation of pond bottoms. The only way to be certain of average pond depth is to measure water depth before treatment at several dozen random locations.

Handle herbicides safely

Although aquatic herbicides are relatively safe to handle, it is nevertheless important to keep exposure to an absolute minimum, especially for applicators. Herbicide labels and material safety data sheets advise what protective clothing and equipment should be worn, any precautions the handler should follow, a statement of practical treatment in case of poisoning, statements concerning hazards to the environment, any physical or chemical hazards, and directions on proper storage and disposal. By law, copies of labels and any supplementary labels must be in the possession of the applicator at the application site for each herbicide used. Anyone who handles a pesticide must read and understand all label statements prior to using the product. Herbicide safety is reviewed by Avery (2003).

Be aware of the consequences of herbicide use

With the notable exceptions of copper-based algicides and the methylamine formulations of endothall, herbicides registered for use in United States aquaculture are of low toxicity to most aquatic animals and, when used at recommended rates, seldom cause direct toxic-ity problems. However, the use of herbicides in ponds has risks because treatment of a plant-infested body of water with any phytotoxic chemical can cause a dramatic change in environmental conditions that may endanger cultured animals. In particular, decay of plants killed by the herbicide may lead to dissolved oxygen depletion. The extent to which water quality is affected depends on the amount of plant material killed, the amount of plant material unaffected by the herbicide, the rate death occurs, water temperature, and other factors. Obviously, the risks of herbicide treatment are greatest when treating dense stands of weeds during hot weather because the dead plant material is rapidly decomposed and the solubility of oxygen in water is low. It is usually inadvisable to use herbicides under such conditions, but if treatment is absolutely necessary, the pond should be divided into halves or thirds and only one portion of the pond treated at one time. After waiting a few days, another portion can be treated, and so on. With this method, there may not be enough vegetation killed at any one time to cause an oxygen depletion. In any case, supplemental aeration should be available to prevent fish kills caused by low dissolved oxygen concentration resulting from plant decomposition.

Dispose of herbicide containers properly

Improper disposal of herbicide containers can cause contamination of soil and water and may result in fi nes or loss of an applicator's license. Empty herbicide containers must be triple rinsed, with each rinsing drained into the herbicide mix tank. If no mix tank is used, the rinse water from the container should be applied to the pond in the same manner as the herbicide in the container. Containers must then be punctured or crushed so that they cannot be reused. Empty bags must be rinsed or shaken clean and cut so that they cannot be used for other purposes. Laws regarding disposal of rinsed containers vary among states, so be sure to follow all state and local regulations regarding pesticide container disposal.

Two aquatic herbicides, 2,4-D and endothall, are regulated as hazardous materials under federal law, and any waste generated during their use must be disposed of as hazardous waste. Triple-rinsed containers can be disposed of as with any other pesticide container. Any rinse water from cleaning of containers or application equipment must be applied as if it were the herbicide or disposed of at a hazardous waste disposal facility.

Bibliography

Bergin, A.: *The Pacific Patrol Boat Project: A Case Study in Australian Defence. Cooperation*, Australian Defence Force Academy, Canberra, 1994.

Boyd, Robert: *People of The Dalles: The Indians of Wascopam Mission*, Lincoln, University of Nebraska Press, 1996.

Bullard, Oral: *Crisis on the Columbia*, Portland, Touchstone Press, 1968.

Cohen, Fay: *Treaties on Trial: The Continuing Controversy over Northwest Indian Fishing Rights*, Seattle. University of Washington Press, 1986.

Cone, Joseph: *A Common Fate: Endangered Salmon and the People of the Pacific Northwest*, Corvallis, Oregon State University Press, 1995.

Crane, Barry D. and Steve Warner, Melissa Hirsch Kuchma: *Fisheries Law Enforcement: Assessment of Deterrence*, Institute for Defense Analyses, Virginia, 1999.

Kelleher, K. and P. Cacaud: *Governance of Yemen's Fisheries: Analysis and Project Financing Proposal*, Prepared for the European Commission, Tebodin, 1999.

Landeen, Dan and Allen Pickham: *Salmon and His People: Fish and Fishing in Nez Perce Culture,* Lewiston, Idaho: Confluence Press, 1999.

Peterson, Keith: *River of Life, Channel of Death: Fish and Dams on the Lower Snake*, Lewiston, Idaho: Confluence Press, 1995.

Richards, A.: *Reported Fisheries Violations in the EEZs of FFA Member Countries, 1978 - 2001: An Analysis of the Violations and Prosecutions Database*, Forum Fisheries Agency, 2001.

Taylor, Joseph: *Making Salmon: An Environmental History of the Northwest Fisheries Crisis*, Seattle, University of Washington Press, 1999.

Ulrich, Roberta: *Empty Nets: Indians, Dams, and the Columbia River*, Corvallis, Oregon State University Press, 1999.

Index

□□□